Service Robotics

*Edited by Volkan Sezer, Sinan Öncü
and Pınar Boyraz Baykas*

Published in London, United Kingdom

IntechOpen

Supporting open minds since 2005

Service Robotics
http://dx.doi.org/10.5772/intechopen.87712
Edited by Volkan Sezer, Sinan Öncü and Pınar Boyraz Baykas

Part of IntechOpen Book Series: Mobile Robotics, Volume 2
Book Series Editor: Efren Gorrostieta Hurtado

Contributors
Khattab M Ali Aheeti, Salah Sleibi Alrawi, Duaa Al_Dosary, Bülent Özkan, Belma Kencebay, Liang Gong, Chengliang Liu, Wei Zhang, Pinar Boyraz, Ertugrul Bayraktar, Cihat Bora Yigit, Anthony Tzes, Nikolaos Evangeliou, Athanasios Tsoukalas, Nikolaos Giakoumidis, Steffen Holter, Smita Nayak, Rajesh Das, Volkan Sezer, Hosein Houshyari, Rahman Salim Zengin, Murat Cenk Yılmaz

First published in London, United Kingdom, 2020 by IntechOpen
IntechOpen is the global imprint of INTECHOPEN LIMITED, registered in England and Wales, registration number: 11086078, 5 Princes Gate Court, London, SW7 2QJ, United Kingdom
Printed in Croatia

British Library Cataloguing-in-Publication Data
A catalogue record for this book is available from the British Library

Additional hard and PDF copies can be obtained from orders@intechopen.com

Service Robotics
Edited by Volkan Sezer, Sinan Öncü and Pınar Boyraz Baykas
p. cm.
Print ISBN 978-1-78984-428-3
Online ISBN 978-1-78984-429-0
eBook (PDF) ISBN 978-1-83968-030-4
ISSN 2632-5195

For EU product safety concerns:
IN TECH d.o.o., Prolaz Marije Krucifikse Kozulić 3, 51000 Rijeka, Croatia, info@intechopen.com or visit our website at intechopen.com.

IntechOpen Book Series
Mobile Robotics
Volume 2

Volkan Sezer is currently an Associate Professor in the Control and Automation Engineering Department of Istanbul Technical University. He received his B.Sc. in Electronics and Telecommunication Engineering from Yildiz Technical University, Istanbul, Turkey in 2005, his M.Sc. in Mechatronics Engineering and his Ph.D. in Control and Automation Engineering from Istanbul Technical University, Istanbul, Turkey in 2008 and 2012 respectively. After his PhD. he did research in Singapore as an MIT (Massachusetts Institute of Technology) researcher in the Future Urban Mobility (FM) group. His research interests are based on the artificial intelligence for robotics and automotive technology. More specifically autonomous/semi-autonomous ground vehicles, active safety, robot autonomy and energy efficiency of hybrid electric vehicles are the main subjects of his research.

Sinan Öncü received his B.Sc. degree in electronics and telecommunications engineering and his M.Sc. degree in mechatronics engineering from Istanbul Technical University (ITU), in 2005 and 2008, respectively, and his Ph.D. degree in mechanical engineering from the Eindhoven University of Technology (TU/e), in 2014. From 2013 to 2016, he was a research scientist at the Integrated Vehicle Safety Department, Netherlands Organization for Applied Scientific Research (TNO) Technical Sciences. Since 2019, he has been an assistant professor in the Mechanical Engineering Department at Boğaziçi University where he is also the Director of Smart and Autonomous Mobility Research Lab. His research focuses on cooperative localization, sensing, control, and coordination of automated air and ground vehicles with a particular focus on automotive and mobile robotics applications.

Pinar Boyraz received double-major B.Sc. degrees in mechanical and textile engineering from Istanbul Technical University (ITU), İstanbul, Turkey, in 2003 and 2004 and a Ph.D. degree in mechatronics from the Wolfson School of Mechanical and Manufacturing Engineering, Loughborough University, Loughborough, U.K., in 2008. From 2008 to 2010, she was a Post-doctoral RA in the Erik Jonsson School of Engineering and Computer Science, University of Texas at Dallas, USA, focusing on driver behavior modeling and active safety system development. From 2010 to 2014, she was an Assistant Professor and from 2014 to 2018, an Associate Professor of the Mechanical Engineering Department, ITU, conducting research in applied robotics. Since March 2018, she has been an Associate Professor with the Mechanics and Maritime Sciences Department, Chalmers University of Technology, Gothenburg, Sweden. Her research interests broadly include applications of mathematical modeling, mechatronics, signal processing, and control theory. She was awarded by the Alexander von Humboldt Foundation with Experienced Researcher Fellowship during her research in applied robotics at Leibniz Universität Hannover, Germany, in 2016 and 2017.

Editors of Volume 2:
Volkan Sezer
Department of Control and Automation Engineering,
Istanbul Technical University, Istanbul, Turkey

Pinar Boyraz Baykas
Department of Mechanics and Maritime Sciences (M2),
Chalmers University of Technology, Gothenburg, Sweden

Sinan Oncu
Department of Mechanical Engineering, Bogazici University, Istanbul, Turkey

Book Series Editor:
Efren Gorrostieta Hurtado
Engineering Faculty of the Autonomous University of Queretaro, Mexico

Scope of the Series

Mobile Robotics is a series devoted to applications of mobile robots and interaction of intelligent systems in different parts of the robot. The topics which this series is exploring are robot control, navigation system, intelligent systems, learning systems, deep learning systems, robot design, and applications in different environments. The series also welcomes research work in the area of mobile robotics development.

Contents

Preface

This book is dedicated to the memory of Dr. Pınar Boyraz Baykas, who passed away in November 2020. I wish to thank her for her work in this book, and for everything else she did in life as a person and a scientist. She will be missed very much, but will always remain in our hearts.

As a formal definition from "International Organization for Standardization", service robots are platforms that perform useful tasks for humans or equipment excluding industrial automation applications. From this perspective, the field of service robotics covers a very wide area including; personal care and assistance, agriculture, logistics, mobility, medical, and defense-oriented robots. All these types of robots may have different levels of autonomy, and they become smarter using artificial intelligence techniques day by day. Parallel to this improvement, the service robotics market has been growing very rapidly, especially during the last few years. With the development of the field of service robotics, the definition of social roles, labor, and jobs will change, some of them will become obsolete and some new ones will be introduced into our lives in the future. Since service robotics contains many types of robots, the variety of problems to be solved is also large. Many popular robotic problems, ranging from mechanism design to simultaneous localization and mapping SLAM, from motion planning to system security, can be examined in this context. This book intends to present the reader with the latest solutions for such research problems and innovative applications of different types of service robots.

In Chapter 1, the authors detail three different enabling approaches for obtaining safe human-robot interactions with two main applications in medical and service robotics. These enablers are mainly joint mechanism designs using variable-stiffness-actuators (VSA), hyper-redundant robotic platforms to improve shape-controllability in narrow channels, and finally a smart-skin design for multi-modal perception of force, stretch, and pressure interactions to improve environmental awareness and therefore intelligence of robotic platforms.

Application of artificial intelligence (AI) and robotics technology has a huge impact in achieving independent mobility and enhances the quality of life for persons with disabilities. Chapter 2 is a survey of the known applications of AI to one of the segments of modern healthcare, namely the prosthetic and orthotic industry and in the field of assistive technology. The contents of this chapter cover the description of basic mathematical tools and technologies from AI that are suitable for the solution of problems arising in this field.

Chapter 3 features both hardware and software design aspects and integration for a heavy-lift drone. The application has interesting end-use as it also integrates a robotic arm on the drone for repair, painting, surveillance and different tasks where interaction forces may or may not be present. The combined robotic platform also details object tracking and localization applications so that the higher-level tasks involving manipulation and surveillance can be implemented.

Chapter 4 considers guidance-based control laws as an alternative to motion planning for autonomous systems. The applicability of the guidance laws to autonomous systems is investigated in a manner similar to the approach for guided munitions. For this purpose, the motion planning of the selected robotic platform, which comprises a robotic arm, tracked land vehicle, and quadrotor is orchestrated in order to move them to predefined target points. Having designed the control systems compatible to the selected guidance laws for the considered systems, the corresponding guidance scheme is constructed. Eventually, the applicability of the method is tested and demonstrated through simulation studies where it is observed that the desired target chase can be made in a successive manner for all cases.

Chapter 5 explains the conversion procedure of a conventional electric wheelchair into an autonomous personal transportation testbed. Autonomous and semi-autonomous wheelchairs are good applications of service robotics. The conversion process includes information about by-wire technology, sensors, computational system and human interface. Besides the conversion procedure, the application of several basic autonomous driving algorithms on the developed testbed such as localization and mapping, are illustrated in this chapter.

Chapter 6 is another interesting intelligent wheelchair application that focuses on the security more than its autonomy. An identification model is achieved for the intelligent platform with high rate of accuracy. This is done through the design of an identification process using Integrated Circuit Metric (ICMetric) technology. The generated process of ICMetrics number is heavily based on magnetometer, gyroscope, and accelerometer sensors of the platform.

Chapter 7 proposes a methodology of software development for the continuous operation of the dual-arm picking robot. The methods used in this chapter are mainly from image processing and software engineering areas. It focuses on developing an intelligent software system to be used in agricultural context for picking operation. A hierarchical control software is framed by means of the "Sense Plan Act" (SPA) paradigm. The Finite State Machine (FSM) based experimental design is provided using the Robot Operating System (ROS) framework in this chapter.

The last chapter discusses some possible outcomes of the inevitable impact of robotization on the societal structure. Undoubtedly, the increasing adaptation of robotics will contribute to the extermination of some work areas while some others will emerge. Moreover, the shift towards increased decentralization and democratization of the means of production also exhibits new potential for new models for welfare distribution and socio-economical roles. A review and collection of opinions from leading figures and experts from diverse fields including science and business are presented on how these new technological developments may affect the job market, economy, and more broadly the society as a whole by critically discussing both the optimistic and pessimistic scenarios.

Dr. Volkan Sezer
Associate Professor,
Department of Control and Automation Engineering,
Istanbul Technical University,
Istanbul, Turkey

Dr. Pinar Boyraz Baykas,
Associate Professor,
Department of Mechanics and Maritime Sciences (M2),
Chalmers University of Technology,
Gothenburg, Sweden

Dr. Sinan Oncu
Associate Professor,
Department of Mechanical Engineering,
Bogazici University,
Istanbul, Turkey

Note from the Publisher

It is with great sadness and regret that we inform the contributing authors and future readers of this book that the co-editor, Dr. Pınar Boyraz Baykas, passed away shortly after finishing the book and before having a chance to see its publication.

We would like to acknowledge her contribution and express our gratitude for her pleasant cooperation with us.

IntechOpen Book Department Team November, 2020.

Chapter 1

Safe Human-Robot Interaction Using Variable Stiffness, Hyper-Redundancy, and Smart Robotic Skins

Pinar Boyraz Baykas, Ertugrul Bayraktar and Cihat Bora Yigit

Abstract

In service robotics, safe human-robot interaction (HRI) is still an open research topic, requiring developments both in hardware and in software as well as their integration. In UMAY[1] and MEDICARE-C[2] projects, we addressed both mechanism design and perception aspects of a framework for safe HRI. Our first focus was to design variable stiffness joints for the robotic neck and arm to enable inherent compliance to protect a human collaborator. We demonstrate the advantages of variable stiffness actuators (VSA) in compliancy, safety, and energy efficiency with applications in exoskeleton and rehabilitation robotics. The variable-stiffness robotic neck mechanism was later scaled down and adopted in the robotic endoscope featuring hyper-redundancy. The hyper-redundant structures are more controllable, having efficient actuation and better feedback. Lastly, a smart robotic skin is introduced to explain the safety support via enhancement of tactile perception. Although it is developed for a hyper-redundant endoscopic robotic platform, the artificial skin can also be integrated in service robotics to provide multimodal tactile feedback. This chapter gives an overview of systems and their integration to attain a safer HRI. We follow a holistic approach for *inherent compliancy* via mechanism design (i.e., variable stiffness), *precise control* (i.e., hyper-redundancy), and *multimodal tactile perception* (i.e., smart robotic-skins).

Keywords: variable-stiffness, hyper-redundancy, tactile feedback, smart-skin

1. Introduction

In medical mechatronics, especially in the minimally invasive surgery (MIS) applications, the design challenge often has multiple sources. However, these

---

[1] UMAY was supported by Young Investigator Award of Istanbul Technical University (2011–2016) to build a humanoid robot for rehabilitation uses on children with Autism Spectrum Disorders (ASD), supporting the co-authors of this chapter for their PhD.

[2] MEDICARE-C was funded by Alexander von Humboldt Foundation during the Experienced Researcher Fellowship project of Dr. Pınar Boyraz.

difficulties can be generally grouped in relation to mechanism/structural design, actuation selection, and perception. All three fields are affected by the fact that the room for passage, navigation, and operation is very limited. In addition to restrictions in the size of the device, the interaction between the medical device and the tissues or blood poses deeper questions regarding the accuracy of the control and safety of the patient. Regarding the robotic endoscopes/catheters, they can be used as diagnosis tasks using deliberate palpation as well. The other two obvious tasks could include safe navigation inside a narrow and torturous channel while providing adequate accuracy and stiffness at the location of the operation. This chapter presents three different novel approaches which can provide feasible solution to the design challenges while improving the safety. First, in order to improve the mechanical structure and mobility, hyper-redundant mechanisms are presented in Section 2 featuring three different module designs, emphasizing the reconfigurability and modularity. Then, for providing the adjustable forces and compliant action, variable actuator mechanisms are visited in Section 3. The last innovation path involves the perception upgrade, detailing on tactile sensors in Section 4. Involving the tactile sensing units in the robotic skin or sheath can help obtain better feedback and more accurate diagnosis and/or provide safer operation when there are obstacles in the path. All these three aspects are summarized in **Figure 1**.

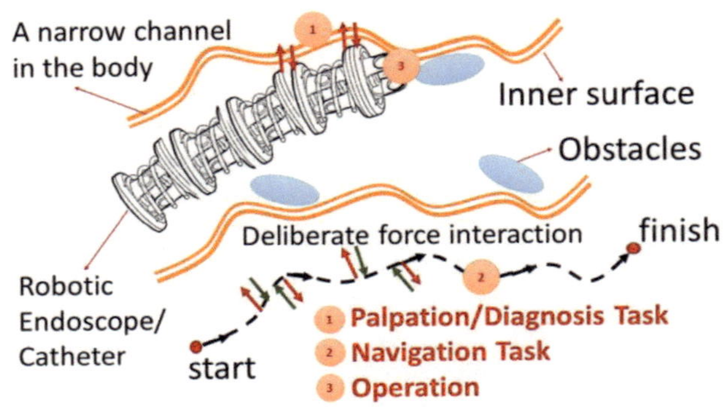

Figure 1.
An endoscopic robotic platform performing diagnosis, navigation, and operation having a hyper-redundant structure, featured in [1, 2].

2. Hyper-redundant mechanisms

Modern surgical robots have been designed and implemented to help surgeons in operations requiring high dexterity and minimal invasiveness. Although great versatility and control have been realized using large, rigid, and serial-link robots such as Da Vinci, catheter-type robotic platforms having smaller dimensions can present an alternative and less expensive solution especially for minimally invasive surgery [3].

In conventional medical use, catheters are manually controlled devices for diagnosis, drug delivery, and basic operations which do not require intricate motion patterns or application of a well-controlled force on the surgical site. Catheters often have a tendon-driven guidewire and a sheath to cover the guiding mechanism which may or may not feature a surgical head/clipper or a micro-mechanism to operate on sensitive tissues. Although originally being passive medical devices, the catheters can be re-designed to gain features such as multiple degrees of freedom, mobility, controllability, and perception. With these improvements in their structure, the catheters can become autonomous or semi-autonomous surgical platforms which can travel in difficult passages in the human body without harming the inner tissues and help the operation itself to be more successful with superior position and force control.

Most of the hyper-redundant or piecewise continuum structures still use rigid or semirigid backbones or general frames. Recently proposed hyper-redundant modular structures can be found in [4–6]. In [4], high dexterity and stiffness requirements are met, whereas the design has poor flexibility, limited compliance, and intricate mechanical structure which can be difficult to miniaturize for medical applications. Other prototypes can feature stiffness control [5] but may fall short in module-based controllability. It is even possible to see applications with better control [6] featuring continuum elastic backbone while segmenting the structure into modules using coupling plates; however, the size and compactness criteria are not fulfilled. The cable-driven structures are lightweight and compact, but there is an inherent limitation of such mechanisms due to cable friction and interdependency between the sections of the modules. As the limitations and drawbacks of such mechanisms are given, in the next section the advantages are highlighted to draw attention to their potential in medical robotics.

2.1 Advantages: modularity and controllability aspects

The main idea of a hyper-redundant robotic platform with a modular building block is to increase the controllability and maneuverability of the robot. Increasing the number of DOF seems to be the main advantage; however, it is surpassed by the fully continuous robot (i.e., tubular/telescopic pre-curved continuum robots) that can be manipulated in 3D space without the need of complicated inverse kinematic calculations due to their inherent compliancy. Although fully continuum robotic platforms have this advantage, for most of the cases, the segment-based local control is not possible to obtain, and in most of the continuum prototypes despite their inherent compliancy, the stiffness control is not possible. Therefore, in this section, we would like to highlight the modular hyper-redundant robotic designs that can offer segment-based position control as well as adjustable stiffness. When the robotic catheter has both the position and the stiffness control, the navigation of the robotic catheter inside the torturous channels becomes an optimal control problem where the position and force are controlled with varying priorities according to the path planning and the operation task. This greatly increases the inherent safety of the robotic catheter. In the following section, designed mechanism modules are introduced and compared using workspace and stiffness analysis.

2.2 Example applications of hyper-redundant mechanisms

A recent hyper-redundant manipulator can be found in [7] which has electromagnetically actuated manipulator. Early examples of hyper-redundant manipulators with full solution on kinematics and dynamics are given in [8–10]. However, control algorithms suggested for such mechanisms are still in progress. For example, a modular control scheme is proposed in [11].

In this section, the design of hyper-redundant and modular robotic structures is detailed by emphasizing the functional properties such as independent module/segment controllability and variable-adjustable stiffness. The proposed designs [12] are aimed at improving both position and force control of such structures employing whole-body shape control and local stiffness control in the robotic catheter.

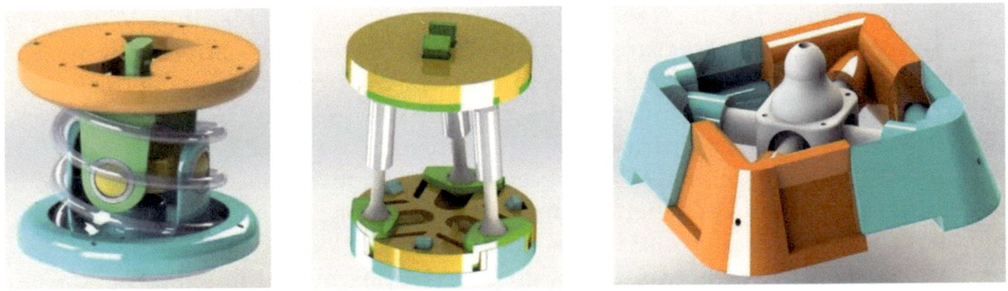

Figure 2.
Modules for hyper-redundant backbone construction, from left to right: Hybrid module, radially reconfigurable 3x SPS parallel kinematic mechanism, and seahorse tail section, featured in [12].

Three different module designs for hyper-redundant mechanisms are depicted in **Figure 2**. The first mechanism is called "hybrid" and essentially a universal joint placed in between two parallel plates and supported by a concentric shaft. This mechanism has three DOF per module, having pan and tilt for adjusting the heading angle while using the translational movement to shrink or elongate while adjusting its stiffness.

The second mechanism is 3x SPS having spherical-prismatic-spherical joints in each strut and is essentially a reconfigurable parallel mechanism. This mechanism provides stiffness adjustments by relocating the connection points of the struts on the lower plate. The struts can also elongate and contract along their axis so that the hyper-redundant platform can be adjusted when passing along difficult cavities.

Finally, the last module is named after the seahorse tail since it is inspired by the cross section of the biological counterpart. This mechanism can radially contract and widen to mimic the function of oblique muscles in seahorse tail structure. The modules are connected by a spherical joint in the middle. Since the radial struts are spring-loaded, the radial stiffness can be adjusted.

The main advantage of hyper-redundant mechanisms is that each segment can be controlled separately, and the multi-degree-of-freedom makes it possible to control the whole-body shape of the manipulator to reduce the risk of harming the tissues during the navigation task. If the modules also have variable stiffness or re-configurability as it is shown here, the versatility and the safety of the hyper-redundant platforms increase. Since robotic platforms should accomplish tasks such as navigation, diagnosis, and operation, they may have to support different levels of stiffness when required. This type of adjustability can be achieved via special actuators as well. The next section expands on this view by supporting these mechanisms with variable stiffness actuators.

3. Variable stiffness actuators

Rehabilitation is known as the process of regaining the deceived somatic talents as a result of an illness or accident, all of which are necessary for survival, quality of life, and living together with their families and society. With the advancement of technological development, specialized mechanisms and devices are used more frequently to resolve some of the issues related with physical interaction between humans and robots. Also these mechanisms or devices operate in clinical environments; some of them are designed to provide mobility for daily usage. Namely, exoskeletons are the wearable types of these mechanisms. Among military usage, civilian purposes, and industrial applications, exoskeletons are for rehabilitation

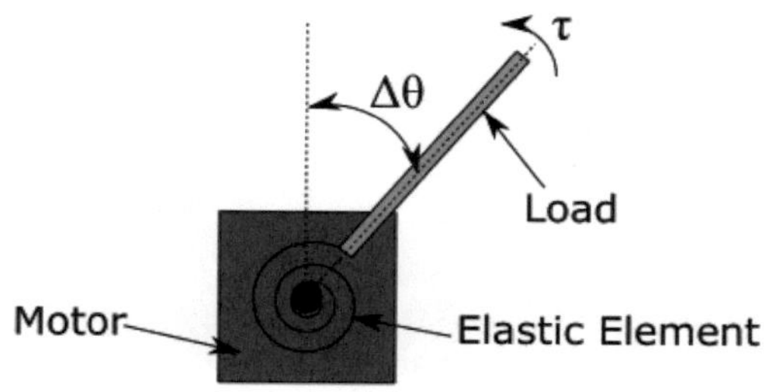

Figure 3.
Schematic representation of series elastic actuators (SEA) or variable stiffness actuators (VSA).

or acquisition of lost actions of people with disabilities. Commonly, similar to other mechatronic systems, exoskeletons comprise sensing, control, and actuation components.

Since rehabilitation is a human-centered therapy to overcome the impairments of the motor functions, it is required to be for exoskeletons to provide safe interaction and mimicking human motions [13]. Compliance is a prerequisite for safety, which can be maintained by software or hardware solutions. Software-based solutions allow controlling impedance [14] by implementing control techniques on rigid joint structures [15]. On the other hand, hardware-based solutions imply flexible joint structures with passive compliance [16, 17]. Also, the interface surface between the patient and the mechanism is covered by soft materials to increase comfort. In addition, adequate amount of force/torque should be supplied to perform the predefined tasks successfully while maintaining a lightweight mechanical structure [18].

Safety is a trending topic within industrial robotic applications. To increase precision, robot joints are designed as stiff as possible; however, Pratt et al. [19] proposed to connect motor and load with elastic components. Consequently, passive compliance is obtained, but it is comprehended that single stiffness value is not suitable for different robotic tasks. Variable stiffness actuators (VSAs) or in general variable impedance actuators (VIAs) are able to adjust the stiffness/impedance within a specific range (see **Figure 3**).

Besides the need of excessive number of human therapists for ordinary rehabilitation techniques, they are time-consuming for labors [20]. Furthermore, these techniques are deficient to measure the performance of rehabilitation outputs for objective analysis. Inflexibility due to different level of treatments, namely, impersonal aims, can be counted as another drawback for traditional rehabilitation therapy methods. In this section, VSA-based exoskeleton/rehabilitation mechanisms are presented as possible solutions to these problems.

3.1 Operational principles of VSAs

Conventional robot joints are designed to track a motion profile and try to keep the position against external effects after reaching the goal position. On the contrary, in the SEA mechanism, there are elastic elements between the load and the motor, which allow the external influences to change in the joint position. The elastic element herein has constant output stiffness (k), and the relation between torque (τ) and position (θ) is linear as follows:

$$k = \frac{d\tau}{d\theta} \tag{1}$$

$$\tau = k\Delta\theta \tag{2}$$

where $\Delta\theta$ denotes the deflection of the elastic element. VSA mechanisms are designed to have nonlinear τ-θ relation yielding variable output stiffness as given in Eq. 3:

$$d\tau = f(\theta)d\theta \tag{3}$$

where $f(\theta)$ is the nonlinear stiffness function. Regarding the change in stiffness, the reaction of the joint for external effects can be adjusted according to the desired task. The stiffness adjustment mechanisms are classified in three main categories in [21]: (i) spring preload, (ii) changing transmission between load and spring, and (iii) physical properties of the spring. The system, in which a couple of springs and motors run in a reciprocal manner, namely, antagonistic springs, is the first type. The working principle resembles biological musculoskeletal system in the first category. To obtain a linear stiffness variation, two quadratic springs are utilized in [22] by using a cam mechanism with a linear helical spring. In [23], the importance of quadratic springs in the design of VSAs is shown. The second type provides nonlinear torque-position relation by changing the distance between rotation center, the linear spring connection points, and/or tip point [24]. The last kind exploits natural characteristics of linear springs. In [25], nonlinearity of helical springs under bending and compression determines stiffness variation. Moreover, mechanism in [26] specifies the number of active coils of helical spring which results a change in stiffness. General schematic representations of the first two types are given in **Figure 4**.

VSAs are generally actuated by conventional electrical motors; however, in [27] stiffness variation is obtained by a pneumatically artificial muscle. Hobby servo motors are another alternative to conventional motors which is used in a modular VSA design to lower the cost [28]. Similar to actuation units, elastic components can vary in different mechanisms. Although it is not implemented in an actuator, a nonlinear spring mechanism in [29] includes rubber, and in [30], a timing belt is introduced as the source of elasticity.

VSAs are superior to conventional actuators according to energy efficiency under various working conditions and performing highly dynamic task. Energy-efficient gait is performed by using compliant actuators because of the energy storage capability of the elastic element. However, when the environment or the walking speed is changed, natural dynamics of the mechanism is expected to maintain efficiency. In [31], running motion energy cost of a legged robot, ***Edubot***, decreases about 40% when it is compared to fixed stiffness legged robots. In addition, dynamic tasks that cannot be accomplished with classical robots can be done by these mechanisms due

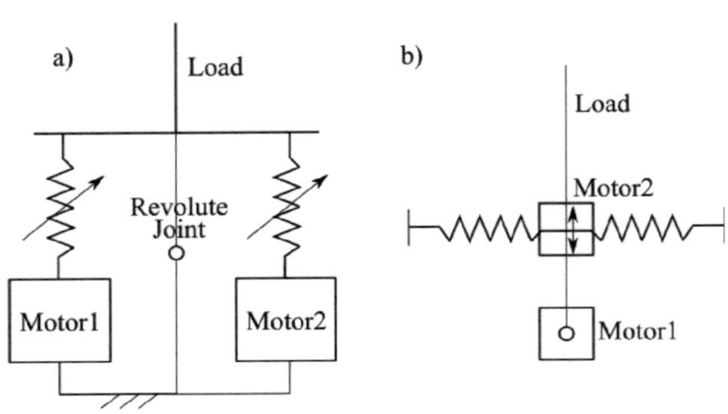

Figure 4.
Stiffness adjustment types: (a) spring preload; (b) changing transmission between load and spring.

to the energy storage feature. An optimal control strategy is implemented to the mechanism to maximize ball-throwing distance in [32]. Benefits of stiffness adjustment is presented in the study by comparing variable and fixed stiffness performances. A detailed analysis of the VSA designs can be found in [21].

3.2 Example applications of VSA

Modern studies towards medical mechatronic systems are performed as interdisciplinary collaboration conducted with physicians, therapists, and scientists among engineering community. Recent approach to these systems serves the emergence of new perspectives beyond the advantages of robot-assisted therapy. Principally, practical studies can be divided into two parts depending on the intention of mechanisms to the upper body and lower body.

Upper body exoskeleton applications are mostly focused on the upper limb and elbow parts. A torque-driven and lightweight exoskeleton called ***Limpact*** is proposed to sustain therapeutic aid for upper limb rehabilitation in [33, 34]. Suitable dimensions for wearable functionality and impairment quantification can be taken into account as further characteristics as well as rotational hydro-elastic actuator as being a new type of SEA. There also exist different control modes such as compliant impedance and stiff admittance. In [16, 17], a 4-DOF wearable passive exoskeleton mechanism for elbow rehabilitation, named as NEUROExos, driven by a variable impedance antagonistic actuator, is presented. The double-shell link structure of NEUROExos contributes to ergonomics, and the joint position and stiffness are controlled separately by passive compliant actuation system. The experiments conducted including a human subject show that the increase in the joint stiffness causes smaller angular error during the motion in the reference trajectory. This is a result of ensuring the proper torque transmission relation between the human and exoskeleton. AVSER [35] is another study towards elbow rehabilitation using an active variable stiffness exoskeleton. Within AVSER, there is an active variable stiffness elastic actuator (AVSEA) composed of two DC motors, one for controlling joint position and the other one is for varying stiffness that is produced by a leaf spring. The effective length of the leaf spring, which is controlled by AVSEA, affects the motion characteristics of AVSER, which can be active or passive. Human-included experiments are conducted using the data gathered from two encoders for motor and elbow angles, a linear potentiometer for linear spring deflection and two active electrodes for electromyogram (EMG) signals. The results display the compatibility of AVSER for active-passive elbow rehabilitation tasks with its capabilities of stiffness adjusting, safety, and energy efficiency.

Lower body assistance can be in forms of full support to the legs, or it can affect only dysfunctional part such as the ankle or knee. Exoskeletons and rehabilitation mechanisms are widely used in the lower limb to regain locomotion of disabled or patients who have difficulty with walking. In addition to gait assistance, standing up motion is provided with the help of exoskeletons. A brace about the foot which is called as ankle-foot orthosis (AFO) is a usual treatment for drop-foot gait. A variable impedance actuator with force and position sensors is assembled to an AFO in [36]. It is shown that during different phases of the gait, adjusting impedance values increases the benefits of AFOs. Sit-to-stand task is a torque demanding task especially for knee joints. In [37], a lever arm mechanism based on VSA for knee exoskeleton is presented, and design methodology is explained in detail. Moreover, the effects of different stiffness values are evaluated for standing task. It is not only necessary to supply sufficient torque to knee joint but also to understand the intention of the user. Instead of splitting task into phases to control stiffness or impedance in [28], muscle activity of the patient is collected via EMG in order to detect patient

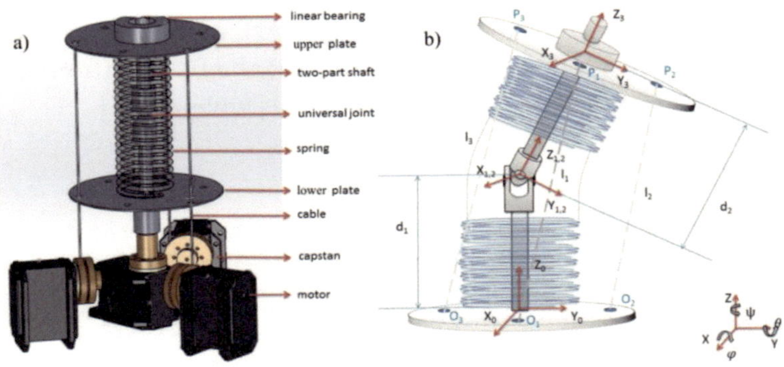

Figure 5.
Variable stiffness neck joint. (a) Components of the mechanism and (b) reduced kinematic model [25].

intention and correct stiffness values in [29]. Along with functional design of VSAs in rehabilitation mechanisms, researchers are inspired from the human muscle structure and designed full lower limb orthosis to improve impaired gait of patients by using pneumatic [38] and wire-based artificial muscles [20]. Furthermore, VSAs are expected to mimic human joint behaviors and have a great potential [39, 40] to be used in rehabilitation purposes. In [25] a VSA which resembles a human neck joint is presented. The schematic representation of the mechanism is given in **Figure 5**. Cable-driven lightweight structure brings simplicity. Also, there is no additional hardware other than a helical spring for stiffness variation. Although the middle shaft restricts the motion due to its revolute-revolute-prismatic (RRP) structure, actuation principle is similar to parallel mechanisms.

All in all, robot-assisted rehabilitation studies and applications are still attractive research areas. Human motion imitation for mechanisms used in rehabilitation is emphasized for successful results. To this end, VSAs are comprised within rehabilitation systems. More information about the latest progress can be found in [41–43]. These findings reveal that robot-assisted technologies will result in less human labor time consumption with increasing quality of observable rehabilitation outputs.

4. Smart robotic skins

For connecting the robot to its environment, visual sensor channels are usually preferred and widely applied. The tactile sensor applications are limited to certain locations, which tends to be the tip of the device especially for robotic catheters. In this part of the chapter, a large-area application for the tactile sensors to form an artificial skin on the robotic catheter is covered. The large-area, skin-like applications of tactile sensors can empower the robotic catheters to have better perception output during diagnosis/palpation while helping to obtain higher safety levels during operations.

4.1 Tactile sensors for medical robotics

In one of the recent surveys on state-of-the-art tactile sensing for minimally invasive surgery (MIS) [44], it is clearly stated that the best place to include sensing elements in MIS device is on the instrument shaft inside the patient's body. The force sensors on the tip of the endoscopic tools are not strongly suggested, because the space is very limited. Incorporating a tip sensor involves either having a larger gripper or manufacturing of extremely small transducers. The general overviews on the tactile sensors without a specific focus on use in the surgical robotics can be seen in [45–48]. In [46], the focus of the overview was extended to electronic skin technologies, whereas in [45] the effective utilization of the tactile skin takes the contact condition into special consideration. Some overviews focus on the wearable features [47], and the others explain the difficulties in the development of tactile sensor units emphasizing its complexity involving multiple transduction ways [48]. In order to develop tactile sleeves/sheaths for MIS endoscopic robotic platforms, a broader perspective of current tactile technology development is needed. Although the application is very different and does not contain any tactile modalities, in [49] a flexible and wearable skin for health monitoring interface is reported. These types of advanced skin patches can even be used for scheduled drug delivery [50]. Some of the relevant studies can be found from soft robotics literature. For example, in [51] a shape-tracking algorithm using polyvinylidene fluoride (PVDF)-based sensors on the hyper-flexible beams is used. Although the beam is in 2D, the proposed method can be extended to 3D providing a spatial ego-motion tracking for flexible endoscopic robots. The research [52] reports a discrete piezo-ceramic sensor array embedded in soft substrate, therefore offering a solution to accuracy problems in film-based piezo material but at the same time providing some compliance and stretchable behavior. Although the piezo-electric transduction is very widely used, there are also alternative methods based on optical modality. For example, in [53], a large-area sensor for pressure measurement was suggested using organic field-effect transistors (OFETs). Similarly, in [54] an optical principle is used to measure data through employing fiber Bragg grating and waveguides inside the compliant substrate material. The waveguide approach is also used in [55] but employing PDMS as the substrate this time.

4.2 Example application of tactile sensors as robotic skin

Herein, an example application is presented from AvH Project, MEDICARE [1], together with the measurement methodology it uses. The manufacturing of the tactile sleeve is achieved using multiple layers of silicone substrate in an additive manner to embed the piezo sensors in the desired depth and location. The silicone substrate was selected as Eco-flex 00-10 because of its relatively easy vacuuming and curing procedures. In addition to these advantages, the mechanical properties of Eco-flex are very close to the human tissue, and it is relatively low-cost. The distance between pressure sensors is large in this setup; however, ideally, they can be arranged with 4 mm separation in each active cell. The data cables connecting the sensors to the data acquisition circuit are soldered carefully, and meandering shapes are given to the bare wires to prevent fractures within the substrate when the sleeve moves with the backbone. It must be stated that using off-the-shelf sensors limits the stretchability of the sensing areas; still, the sleeve remains flexible enough to be wrapped around a robotic backbone. The tactile sleeve is produced in a flat sheet (**Figure 6a**) having slanted edges and was connected to the backbone in cylindrical form (**Figure 6b**) in the second step. The slanted angles at the edges allowed connecting the sleeve without having a bulk on the connection line. As it can be seen in **Figure 6**, the silicone sleeve features a ripple structure on the outer surface. This structure is a first attempt to increase the perception capacity of embedded sensors using the structural computation.

The measurement methodology of the sleeve when the outer surface of the silicone sleeve contacts with a rough surface, the ripples would help create a high-frequency interpretation of the surface properties in the sensor output. Although

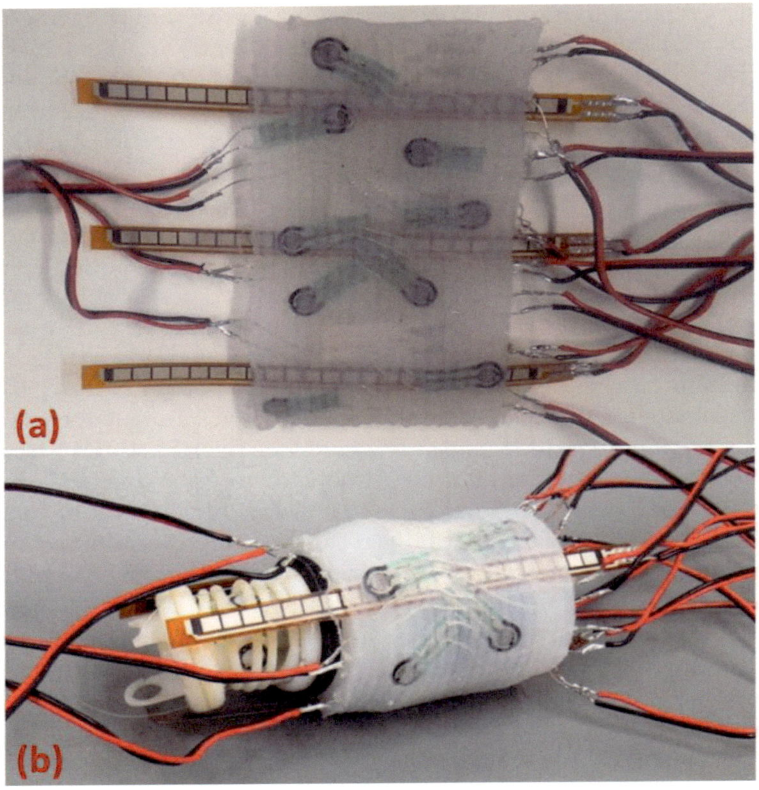

Figure 6.
Robotic sheath for endoscopic hyper-redundant platform developed in AvH project by Dr. Boyraz at Leibniz University of Hannover: (a) flat; (b) wrapped around the robotic platform, featured in [2].

being very simple, the surface ripple structures can be elaborated to include multiscale ripples in a fractal manner to interpret different surface structures having different frequency in the vibration pattern.

5. Conclusion

In this chapter, three different enabling technologies at research frontiers have been discussed to support safe interactions between robots and humans in different fields of robotics, mainly focusing on service, medical, and rehabilitation areas. We provided an extensive overview of variable stiffness, hyper-redundancy, and smart-skin structures in application cases. All these structures are only some of the enablers for safe HRI, and future studies should concentrate on integrating them in service robotics for the full benefit. While the variable stiffness offers inherent compliance during interaction and adaptability of the structures, the hyper-redundancy may allow better controllability of available degrees of freedom. Finally, the smart robotic skins can provide crucial feedback for safer interactions. As a holistic approach to safe HRI, inherent or structural compliance, enhanced controllability, and improved tactile feedback can bring significant safety built in robotic structures. In future work, the robotics platform that can integrate all of them may even have synergistic effects of the combined subsystems as the tactile feedback may directly be linked to variable stiffness adjustment or reconfigure the active links in a hyper-redundant structure.

Acknowledgements

This work has been partially funded by Alexander von Humboldt Foundation during the Experienced Researcher Fellowship project of Dr. Pinar Boyraz at Leibniz University of Hannover under the supervision of Prof. Tobias Ortmaier and Prof. Annika Raatz.

Author details

Pinar Boyraz Baykas[1*†], Ertugrul Bayraktar[2] and Cihat Bora Yigit[3]

1 Department of Mechanics and Maritime Sciences (M2), Chalmers University of Technology, Gothenburg, Sweden

2 Visual Geometry and Modelling (VGM) Lab, Istituto Italiano di Tecnologia, Genova, Italy

3 Siemens A.S., Digital Industry Division, Motion Control-Machine Tool Systems, Kartal, Istanbul, Turkey

*Address all correspondence to: boyraz.pinar@googlemail.com

† Deceased.

References

[1] Available from: AvH Project website: https://cmsv050i.t3luh.uni-hannover.de/?id=296 [Accessed: 05 January 2018]

[2] Boyraz P, Tappe S, Ortmaier T, Raatz A. Design of a low-cost tactile robotic sleeve for autonomous endoscopes and catheters. Measurement and Control. 2020. DOI: 10.1177/0020294019895303

[3] Huda MN, Yu H, Cang S. Robots for minimally invasive diagnosis and intervention. Robotics and Computer-Integrated Manufacturing. 2016;**41**:127-144

[4] Salomon O, Wolf A. Inclined links hyper-redundant elephant-trunk-like robot. Journal of Mechanics and Robotics. 2012;**4**:6

[5] Li Z, Ren H, Chiu PWY, Du R, Yu H. A novel constrained wire-driven flexible mechanism and its kinematic analysis. Mechanism and Machine Theory. 2016;**95**:59-75

[6] Tonapi MM, Godage IS, Vijaykumar AM, Walker ID. A novel continuum robotic cable aimed at applications in space. Advanced Robotics. 2015;**29**(6):861-875

[7] Tappe S, Dorbaum M, Kotlarski J, et al. Kinematics and dynamics identification of a hyper-redundant, electromagnetically actuated manipulator. In: Book Series IEEE/ASME International Conference on Advanced Intelligent Mechatronics (AIM); 12-15 July 2016

[8] Chirikjian GS, Burdick JW. A modal approach to hyper-redundant manipulator kinematics. IEEE Transactions on Robotics and Automation. June 1994;**10**(3):343-354

[9] Chrikjian GS, Burdick JW. Kinematically optimal hyper-redundant manipulator configurations. IEEE Transactions on Robotics and Automation. 1995;**11**(6):794-806

[10] Chrikjian GS. Hyper-redundant manipulator dynamics: A continuum approximation. Advanced Robotics. 1994;**9**(3):217-243

[11] Cho NC, Jung H, Son J, Kim KG. A modular control scheme for hyper-redundant robots. International Journal of Advanced Robotic Systems. 2015;**12**:91

[12] Nastar PR, Boyraz P, Ortmaier T, Raatz A. Development of compliant hyper-redundant mechanisms for robotic catheters and analysis of controllability. In: DGR Days in Connection with Robocup, Leipzig, Germany. June 2016

[13] Riener R, Lünenburger L, Colombo G. Human-centered robotics applied to gait training and assessment. Journal of Rehabilitation Research and Development. 2006;**43**(5):679-694. DOI: 10.1682/JRRD.2005.02.0046

[14] Hogan NIC. An approach to manipulation. In: American Control Conference (ACC); 6-8 June 1984; San Diego. New York: IEEE. 1984. pp. 304-313

[15] Aguirre-Ollinger G, Colgate JE, Peshkin MA, Goswami A. Active-impedance control of a lower-limb assistive exoskeleton. In: IEEE 10th International Conference on Rehabilitation Robotics (ICORR); 13-15 June 2007; Noordwijk. New York: IEEE. 1984. pp. 188-195

[16] Lenzi T, Vitiello N, De Rossi SMM, Roccella S, Vecchi F, Carrozza MC. NEUROExos: A variable impedance powered elbow exoskeleton. In: IEEE International Conference on Robotics and Automation (ICRA); 9-13 May 2011; Shanghai. New York: IEEE. 2011. pp. 1419-1426

[17] Vitiello N, Lenzi T, Roccella S, De Rossi SMM, Cattin E, Giovacchini F, et al. NEUROExos: A powered elbow exoskeleton for physical rehabilitation. IEEE Transactions on Robotics. 2013;**29**:220-235. DOI: 10.1109/TRO.2012.2211492

[18] Tsagarakis NG, Caldwell DG. Development and control of a 'soft-actuated' exoskeleton for use in physiotherapy and training. Autonomous Robots. 2003;**15**:21-33. DOI: 10.1023/A:1024484615192

[19] Pratt GA, Williamson MM. Series elastic actuators. In: Proceedings of the IEEE/RSJ International Conference on Intelligent Robots and Systems 'Human Robot Interaction and Cooperative Robots'; 5-9 August 1995; Pittsburgh. New York: IEEE. 1995. pp. 399-406

[20] Surdilovic D, Zhang J, Bernhardt R. STRING-MAN: Wire-robot technology for safe, flexible and human-friendly gait rehabilitation. In: IEEE 10th International Conference on Rehabilitation Robotics (ICORR); 13-15 June 2007; Noordwijk. New York: IEEE. 2007. pp. 446-453

[21] Vanderborght B, Albu-Schäffer A, Bicchi A, Burdet E, Caldwell DG, Carloni R, et al. Variable impedance actuators: A review. Robotics and Autonomous Systems. 2013;**61**(12):1601-1614. DOI: 10.1016/j.robot.2013.06.009

[22] Migliore SA, Brown EA, DeWeerth SP. Biologically inspired joint stiffness control. In: IEEE International Conference on Robotics and Automation (ICRA); 18-22 April 2005; Barcelona. New York: IEEE. 2005. pp. 4508-4513

[23] English CE, Russell D. Mechanics and stiffness limitations of a variable stiffness actuator for use in prosthetic limbs. Mechanism and Machine Theory. 1999;**34**(1):7-25. DOI: 10.1016/S0094-114X(98)00026-3

[24] Jafari A, Tsagarakis NG, Caldwell DG. AwAS-II: A new actuator with adjustable stiffness based on the novel principle of adaptable pivot point and variable lever ratio. In: IEEE International Conference on Robotics and Automation (ICRA); 9-13 May 2011; Shanghai. New York: IEEE. 2011. pp. 4638-4643

[25] Yigit CB, Boyraz P. Design and modelling of a cable-driven parallel-series hybrid variable stiffness joint mechanism for robotics. Mechanical Sciences. 2017;**8**(1):65. DOI: 10.5194/ms-8-65-2017

[26] Hollander KW, Sugar TG, Herring DE. Adjustable robotic tendon using a 'Jack spring'. In: 9th International Conference on Rehabilitation Robotics (ICORR); 28 June-1 July 2005; Chicago. New York: IEEE. 2005. pp. 113-118

[27] Sentis L, García JG, Fernández BR, Gonzales M, Paine N. Design, construction and control of a fluidic robotic joint for compliant legged locomotion. In: IEEE International Symposium on Industrial Electronics (ISIE); 27-30 June 2011; Gdansk. New York: IEEE. 2011. pp. 887-894

[28] Catalano MG, Grioli G, Garabini M, Bonomo F, Mancini M, Tsagarakis N, et al. Vsa-cubebot: A modular variable stiffness platform for multiple degrees of freedom robots. In: IEEE International Conference on Robotics and Automation (ICRA); 9-13 May 2011; Shanghai, New York: IEEE. 2011. pp. 5090-5095

[29] Schepelmann A, Geberth KA, Geyer H. Compact nonlinear springs with user defined torque-deflection profiles for series elastic actuators. In: IEEE International Conference on Robotics and Automation (ICRA); 31 May-7 June 2014; Hong Kong. New York: IEEE. 2014. pp. 3411-3416

[30] Tonietti G, Schiavi R, Bicchi A. Design and control of a variable stiffness

actuator for safe and fast physical human/robot interaction. In: IEEE International Conference on Robotics and Automation (ICRA); 18-22 April 2005; Barcelona. New York: IEEE. 2006. pp. 526-531

[31] Galloway KC, Clark JE, Koditschek DE. Variable stiffness legs for robust, efficient, and stable dynamic running. Journal of Mechanisms and Robotics. 2013;**5**(1):011009. DOI: 10.1115/1.4007843

[32] Braun D, Howard M, Vijayakumar S. Optimal variable stiffness control: Formulation and application to explosive movement tasks. Autonomous Robots. 2012;**33**(3):237-353. DOI: 10.1007/s10514-012-9302-3

[33] Stienen AHA, Hekman EEG, ter Braak H, Aalsma AMM, van der Helm FCT, van der Kooij H. Design of a rotational hydroelastic actuator for a powered exoskeleton for upper limb rehabilitation. IEEE Transactions on Biomedical Engineering. 2009;**57**:728-735. DOI: 10.1109/TBME.2009.2018628

[34] Stienen AHA, Hekman EEG, ter Braak H, Aalsma AMM, van der Helm FCT, van der Kooij H. Design of a rotational hydro-elastic actuator for an active upper-extremity rehabilitation exoskeleton. In: 2nd IEEE RAS & EMBS International Conference on Biomedical Robotics and Biomechatronics (BioRob); 19-22 Oct. 2008; Scottsdale. New York: IEEE. 2008. pp. 881-888

[35] Wang RJ, Huang HP. AVSER—Active variable stiffness exoskeleton robot system: Design and application for safe active-passive elbow rehabilitation. In: IEEE/ASME International Conference on Advanced Intelligent Mechatronics (AIM); 11-14 July 2012; Kachsiung, New York: IEEE. 2012. pp. 220-225

[36] Blaya JA, Herr H. Adaptive control of a variable-impedance ankle-foot orthosis to assist drop-foot gait. IEEE Transactions on Neural Systems and Rehabilitation Engineering. 2004;**12**(1):24-31. DOI: 10.1109/TBME.2009.2018628

[37] Karavas NC, Tsagarakis NG, Caldwell DG. Design, modeling and control of a series elastic actuator for an assistive knee exoskeleton. In: 4th IEEE RAS & EMBS International Conference on Biomedical Robotics and Biomechatronics (BioRob); 24-27 June 2012; Rome. New York: IEEE. 2012. pp. 1813-1819

[38] Costa N, Caldwell DG. Control of a biomimetic "soft-actuated" 10dof lower body exoskeleton. In: The First IEEE/RAS-EMBS International Conference on Biomedical Robotics and Biomechatronics, (BioRob); 20-22 February 2006; Pisa. New York: IEEE. 2006. pp. 495-501

[39] Cestari M, Sanz-Merodio D, Arevalo JC, Garcia E. An adjustable compliant joint for lower-limb exoskeletons. IEEE/ASME Transactions on Mechatronics. 2015;**20**(2):889-898. DOI: 10.1109/TMECH.2014.2324036

[40] Karavas N, Ajoudani A, Tsagarakis N, Saglia J, Bicchi A, Caldwell D. Tele-impedance based assistive control for a compliant knee exoskeleton. Robotics and Autonomous Systems. 2015;**73**:78-90. DOI: 10.1016/j.robot.2014.09.027

[41] Dollar AM, Herr H. Lower extremity exoskeletons and active orthoses: Challenges and state-of-the-art. IEEE Transactions on robotics. 2008;**24**(1):144-158. DOI: 10.1109/TRO.2008.915453

[42] Viteckova S, Kutilek P, Jirina M. Wearable lower limb robotics: A review. Biocybernetics and Biomedical Engineering. 2013;**33**(2):96-105. DOI: 10.1016/j.bbe.2013.03.005

[43] Jimenez-Fabian R, Verlinden O. Review of control algorithms for robotic ankle systems in lower-limb orthoses, prostheses, and exoskeletons. Medical Engineering & Physics. 2012;**34**(4):397-408. DOI: 10.1016/j.medengphy.2011.11.018

[44] Puangmali P, Althoefer K, Seneviratne LD, Murphy D, Dasgupta P. State-of-the-art in force and tactile sensing for minimally invasive surgery. IEEE Sensors Journal. 2008;**8**(4):371-381

[45] Dahiya RS, Mittendorfer P, Valle M, Cheng G, Lumelsky VJ. Directions toward effective utilization of tactile skin: A review. IEEE Sensors Journal. 2013;**13**(11):4121-4138

[46] Seminara L, Pinna L, Ibrahim A, Noli L, Capurro M, Caviglia S, et al. Electronic skin: Achievements, issues and trends. Procedia Technology. 2014;**15**:549-558

[47] Yang T, Xie D, Li Z, Zhu H. Recent advances in wearable tactile sensors: Materials, sensing mechanisms and device performance. Materials Science and Engineering. 2017;**115**:1-37

[48] Lee MH, Nicholls HR. Tactile sensing for mechatronics—A state of the art survey. Mechatronics. 1999;**9**:1-31

[49] Gao W, Enaninejad S, Nyein HYY, Challa S, et al. Fully integrated wearable sensor arrays for multiplexed in situ perspiration analysis. Nature. January 2016;**529**:509-526

[50] Lee H, Choi TK, Lee YB, Cho HR, Ghaffari R, Wang L, et al. A graphene-based electrochemical device with thermos-responsive microneedles for diabetes monitoring and therapy. Nature Nanotechnology. 2016. DOI: 10.1038/nnano.2016.38

[51] Shapiro Y, Kosa G, Wolf A. Shape tracking of planar hyper-flexible beams via embedded PVDF deflection sensors. IEEE/ASME Transactions on Mechatronics. 2014;**19**(4):1260-1267

[52] Acer M, Salerno M, Agbeviade K, Paik J. Development and characterization of silicone embedded distributed piezo-electric sensors for contact detection. Smart Material Structure. 2015;**24**(15):075030

[53] Someya T, Sekitani T, Iba S, Kato Y, Kawaguchi H, Sakurai T. A large-area, flexible pressure sensor matrix with organic field-effect transistors for artificial skin applications. Proceedings of the National Academy of Sciences of the United States of America. 2004;**101**(27):9966-9970

[54] Missinne J, Hoe BV. Artificial skin based on flexible optical tactile sensors. SPIE Newsroom. 2010. DOI: 10.1117/2.1201001.002582

[55] Ramuz M, Tee BC-K, Tok JB-H, Bao Z. Transparent, optical, pressure-sensitive artificial skin for large-area stretchable electronics. Advanced Materials. 2012;**24**:3223-3227

Chapter 2

Application of Artificial Intelligence (AI) in Prosthetic and Orthotic Rehabilitation

Smita Nayak and Rajesh Kumar Das

Abstract

Technological integration of Artificial Intelligence (AI) and machine learning in the Prosthetic and Orthotic industry and in the field of assistive technology has become boon for the Persons with Disabilities. The concept of neural network has been used by the leading manufacturers of rehabilitation aids for simulating various anatomical and biomechanical functions of the lost parts of the human body. The involvement of human interaction with various agents' i.e. electronic circuitry, software, robotics, etc. has made a revolutionary impact in the rehabilitation field to develop devices like Bionic leg, mind or thought control prosthesis and exoskeletons. Application of Artificial Intelligence and robotics technology has a huge impact in achieving independent mobility and enhances the quality of life in Persons with Disabilities (PwDs).

Keywords: artificial neural network, deep learning, brain computer Interface (BCI), electromyography (EMG), electroencephalogram (EEG)

1. Introduction

Human is the most intelligent creature in the planet for their brain power and neural network. The human brain is extremely complex with more than 80 billion neurons and trillion of connections [1]. Simulation scales can array from molecular and genetic expressions to compartment models of subcellular volumes and individual neurons to local networks and system models [2]. Deep Neural Network nodes are an over simplification of how brain synapses work. Signal transmission in the brain is dominated by chemical synapses, which release chemical substances and neurotransmitters to convert electrical signals via voltage-gated ion channels at the presynaptic cleft into post-synaptic activity. The type of neurotransmitter characterizes whether a synapse facilitates signal transmission (excitatory role) or prevents it (inhibitory role). Currently, there are tenths of known neurotransmitters, whereas new ones continuously emerge with varying functional roles. Furthermore, dynamic synaptic adaptations, which affect the strength of a synapse, occur in response to the frequency and magnitude of the presynaptic signal and reflect complex learning/memory functions, (Spike time dependent plasticity) [3, 4]. Recently, evidence has found that surrounding cells, such as glia cells that are primarily involved in 'feeding' the neurons, can also affect their function via the release of neurotransmitters. This new vision of "tripartite synapses," composed of perisynaptic glia in addition to pre- and postsynaptic terminals certainly makes this one of the most exciting discoveries in current neurobiology [5].

The functional loss due to amputation, spinal cord injury, brachial plexus injury or traumatic brain injury resulting loss of connection from brain to extremity and those residual/weakened extremities are not able to function as of healthy/intact limb. These lost structure & functions of extremities were being replaced by fitment of prosthetics and orthotic devices or rehabilitation aids. The conventional prosthesis which is a mechanical device only provide the basic function, similarly Orthosis provides the support to weaken parts not fully with out completely mimicking the lost section. The concept of biomechatronic is a sub-discipline of mechatronics. It is related to develop mechatronics systems which assist or restore to human body gave the prosthetics and orthotics concept to a new direction. A biomechatronic system has four units: Biosensors, Mechanical Sensors, Controller, and Actuator [6]. Biosensors detect intentions of human using biological reactions coming from nervous or muscle system. The controller acts as a translator among biological and electronic structures, and also monitors the activities of the biomechatronic device. Mechanical sensors measure data about the biomechatronic device and relay to the biosensor or controller. The actuator is an artificial muscle (robot mechanism) that produces force or movement to aid or replace native human body function. The areas of use of biomechatronic are orthotics, prosthesis, exoskeleton and rehabilitation robots, and neuroprosthesis. Robots are the intelligent devices that easily fulfill the requirements of cyclic movements in rehabilitation, better control over introduced forces; accurately reproduce required forces in repetitive exercises and more precise in different situations [7].

2. History of artificial intelligence (AI) in prosthetics and orthotics

The first intelligent prosthesis developed by Chas. A. Blatchford & Sons, Ltd. in 1993 [8] and the improved version in 1995 named as Intelligent Prosthesis Plus [9] Blatchford in 1998 developed Adaptive prosthesis combining three actuation mechanisms of hydraulic, pneumatics and microprocessor. The fully microprocessor control knee developed in 1997 by Ottobock known as C-leg [10]. Rheo knee and power knee both developed by OSSUR in 2005 and 2006 subsequently uses onboard AI mechanism [11]. In late 2011 Ossur introduced the world first bionic leg with robotics mechanism known as "symbionic leg" and this time period the Genium X3 was lunched by Ottobock which allow backward walking and provide intuitive and natural motion during gait cycle [12]. On 2015 Blatchford group introduced Linx the world's first fully integrated limb has seven sensor and four CPU throughout the body of Leg. It allows coordination and synchronization of knee and ankle joint by sensing and analyzing data on user movement, activities, environment and terrain making standing up or walking on ramp more natural. The iwalk BiOM is the world first bionic foot with calf system commercially available from 2011 developed by Dr. Hugh Herr uses robotics mechanism to replicate the function of muscle and tendon with proprietary algorithm [13, 14]. The commercially available microprocessor control foot are Meridium (OttoBock, Germany), Elan (Blatchford, UK), Pro-prio (Össur, Iceland), Triton Smart Ankle (hereinafter referred as TSA) (Otto Bock, Germany), and Raize (Fil-lauer, USA) etc. available from 2011 in the market [15].

The first commercially available bionic hand lunched by Touch bionics in 2007 with individually powered digits and thumb has a choice of grip. The design again embedded with rotating thumb known as i- limb ultra and i- limb revolution designs implanted with Biosim and My i- limb app [16]. Bebionic was commercially available in the market in 2010 manufactured by RSL steeper and lunched by World congress, in 2017 it owned by Ottobock. Bebionic3 allows 14 different hold with two thumb position [17]. Michelangelo hand is the fully articulated robotic hand with electronically actuated thumb first fitted in the year 2010 developed by Ottobock [18]. The concept of brain computer interface (BCI) implemented neuroprosthesis or mind control prosthesis which can able to recognize the real time data and a gadget to get nearly normal function is the demand of the day. The EEG based mind controlled smart prosthetic arm was presented in 2016 IEEE conference but till now this concept is not commercialized [19]. Researchers are on the path of developing more complex devices that mimic the natural brain by implementing artificial intelligence to on board computer that read and reply the nerve signal that transmitted to robotic prosthesis and Orthosis which enhance the function of amputated and paralyzed part of the body.

3. Basic concept of AI and machine learning (ML)

3.1 Machine learning

Machine learning contains elements of mathematics, statistics, and computer science, which is helping to drive advances in the development of artificial intelligence. It is the study of computer algorithms which expands and develops through experiences. This is a subset of AI as shown in **Figure 1**. The ML algorithm methods generally categorized two types supervised and unsupervised learning [20, 21].

3.1.1 Supervised learning

The method of predicting a model on a trained range of inputs learning function to maps the known output, which discover the pattern of new sets of data.

Example 1: To predict the model for microprocessor knee joint which is trained with numerous input or labeled data of the knee angle variation in different sub phase of gait cycle and apply on new amputee to predict the new data by the phase dependent pattern recognition approach.

Example 2: Intuitive myoelectric prosthesis or pattern recognition control prosthesis, FES.

Pattern recognition is an automatically recognition of pattern applied in data analysis, signal processing etc. when the pattern of algorithm trained from labeled data that is supervised learning. When the model of algorithm is fruitfully trained, the model can be used for the prediction of a new data. The ultimate goal of this ML is to develop a successful predictor function. The models of discrete or categorical categories of dependent variables are known as classification algorithm and with continuous value known as regression algorithm. Three basic steps followed to finalize a model are training, validating and application of algorithm to new data. Algorithm used for supervised learning are support vector machines, linear regression, linear discriminant analysis (LDA) etc. This is error based learning.

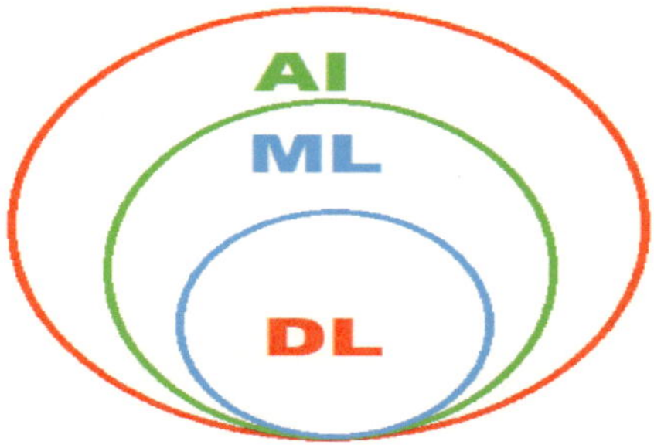

Figure 1.
Relationship between artificial intelligence (AI), machine learning (ML) and deep learning (DL).

Example: Prediction of a model to relate the patient's energy consumption using Trans femoral prosthesis with the function of walking velocity in level surfaces.

The linear regression model for the above statement is:

$$Y = b + aX + e \tag{1}$$

Y = Energy consumption (dependent variable)
b = Y intercept
a = Slope of the Line
X = Walking velocity (Independent variable), e = Error

The Logistic regression model is used to model the probability of a certain class or event such as pass fail, win/loss, healthy/sick etc. This is fall between 0 and 1 with categorical dependent variables.

Example: To predict a model for the successful or failed prosthetic rehabilitation within the categories of 50 meter walk test in level surface with combatable use of any assistive devices for successful and considered as fail if they could not complete the 50 meter walk test.

The model is predicted in terms of the probability (p) which are passing the 50 meter test are pass and could not cross 50 meter as fail.

$$\text{The model of this Statement is}: \ln\left(\frac{p}{1-p}\right) = a + bX \tag{2}$$

p = No of patient cross the level of 50 meter
$1\text{-}p$ = No of patient could not able to cross
The dependent variable Y (predictive) = $p/(1\text{-}p)$
Independent Variable X = Type of prosthesis

3.1.2 Unsupervised learning

The algorithm of unsupervised learning finds a solution to unknown or unlabeled data which is not required any kind of supervision from human. It works of its own to gather information and allow performing more complex task compared to supervised learning. Cluster analysis and k means are the methods used for pattern formation for the new data.

Example: Intent detection algorithm with unlabeled data based on reference pattern is an unsupervised learning method used in microprocessor knee.

3.1.3 Reinforcement learning (RL)

This is concerned with how a software agent must take action in an environment to maximize the cumulative reward. The agent learns from the consequences of its actions and selects the choice from its past experiences and the new choices by the trial and error learning. This is generally output based learning. The components of the RL are agent and environment. The agent (Learner) learns about a policy (π) (strategy or approach that the agent uses to determine the next action based on the current state) by observing or interacting with the environment. All the possible steps followed by the agent during the process of learning are known as the "action" and current condition returned by the environment is "state". The approach that

DOI: http://dx.doi.org/10.5772/intechopen.93903

the agent uses to determine the next action based on the current state is known as "policy". The artificial intelligence gets either reward or penalties for the action the agent performs. The reward is an instant return from the environment to appraise the last action. The goal of an agent to maximize the reward based on the set of actions. The agent follows the concept of exploration and exploitation to get the optimal action value or rewards. The exploration is about exploring and capturing more information from the environment and exploitation uses the already known information to get the reward.

Example: Learning from demonstration (LfD) of myoelectric prosthesis. In this method the policy to determine the next action is learned by different methods i.e. demonstration provided by the Prosthetist, learned from the action of similar prosthetic user or intact limb movement of prosthetic user. During process of demonstration the sequence of state action pairs are recorded for the training of prosthetic limb. The learning process for movement of amputated side with intact limb happens simultaneously. The intact limb considered as training limb and the amputated side prosthetic limb as control limb. During training procedure the agent or learner or amputee asked to perform same motion for both the limb the information from training limb create a prosthetic policy that map the state of action of the control limb. Robotic prosthesis can use its learned and state conditional policy for user during post training use. The training arm demonstrated the desired movement, position and grasp pattern to robotic or control arm. During initial training process the opening of the prosthetic arm may not be the similar to the training limb but when the training preceded the gradual opening of the hand work as a reward to the agent to pick up the appropriate movement and position for required opening of the prosthetic hand and proportional control for graded prehension. The schematic diagram of Bento arm using reinforcement learning shown in **Figure 2** [22]. Another example to understand the strategy of exploring and exploitation is to find out the exact position for placement of surface electrode in the residual limb of amputee. This is a trial and error method where surface electrodes are placed in different locations around the residual limb of the amputee to get the desired action potential to operate the prosthetic hand. The simultaneous activities of residual

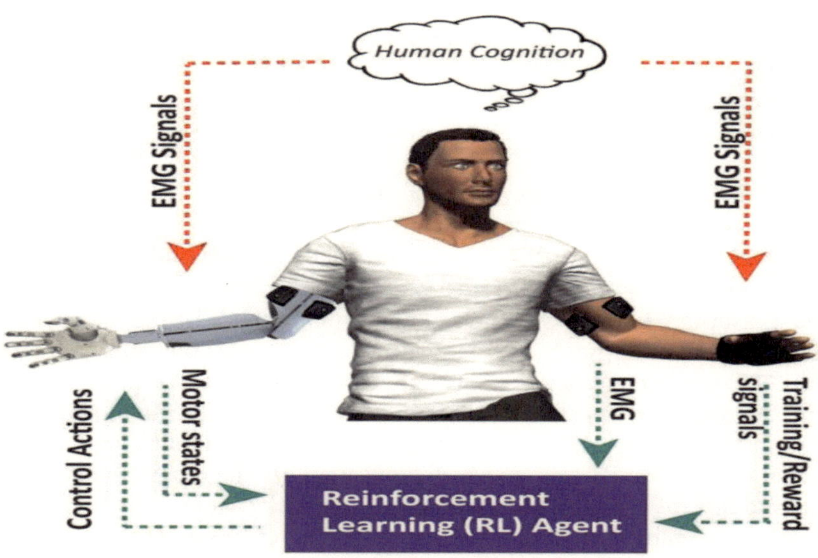

Figure 2.
Schematic diagram of flow of information with bento arm [22].

muscle EMG signal and operation of connected Prosthetic hand provide a visual feedback to amputee and Prosthetist. Based on the feedback the Prosthetist keeps on exploring new site of the electrode in the residual limb until optimization is achieved. This technique helps the amputee to learn about the amount of muscle contraction which operates the prosthesis. The opening and different grasping pattern in sequence acts as a reward to perform more complex activities. In some cases many old user or experienced Prosthetist use the strategy of the exploitation rather than exploring the new site for electrode placement based on their past learning and experiences. Other examples are adaptive switch control myoelectric prosthesis, Power leg Prosthesis, etc.

3.2 Deep learning

This is a form of machine learning uses both supervised and unsupervised and subset of machine learning and AI. It uses the method of artificial neural network (ANN) with representation learning. ANN is inspired by the human brain neural network system whether human brain network is dynamic (Plastic) and analog at the same time the ANN is static and symbolic. It can learn, memorize, generalized and prompted modeling of biological neural system. ANNs are more effective to solve problems related to pattern recognition and matching, clustering and classification. The ANN consist of standard three layer input, output and hidden layer, the output layer can be the input layer for the next output the simple network of neural system shown in **Figure 3** [23], if there many hidden layer are present that ANN known as Deep Neural Networks", or briefly DNN, can be successfully expert to solve difficult problems. Deep learning models yield results more quickly than standard machine learning approaches. The propagation of function in ANN through input layer to output layer and the mathematical representation for this is:

$$s = f\left(\varphi\left(w, x\right)\right) \tag{3}$$

(s = output, x = Input, w = corresponding weight of link between input and transfer function, $\varphi(w, x)$ = linear combination of w and x, f (.) = transfer function.)

Example: EEG based pattern recognition which uses brain computer Interface (BCI) to control prosthetic arm, Neuroprosthesis etc.

3.3 Other artificial intelligence (AI) techniques

Artificial Intelligence is the intelligence of machine that simulates the human intelligence which programmed in such way that it thinks and act like human. It includes; reasoning, knowledge representation, planning, learning, natural language processing, perception, the ability to move and manipulate objects and many more subjects. AI has four main components Expert systems, Heuristic problem solving, Natural Language Processing (NLP) and Vision. In human the intelligent agents like eyes, ears, and other organs act as sensors, and hands, legs, mouth, and other body parts act as per instruction known as effectors similarly the robotic agent substitutes cameras and infrared range finders for the sensors and various motors for the effectors. A software agent has encoded bit strings as its precepts and actions. Similarity between human and artificial intelligence is shown in **Table 1**. AI can be divided into two categories as per its function as symbolic learning (SL) and machine learning (ML). SL is perform the functions like image processing through computer vision

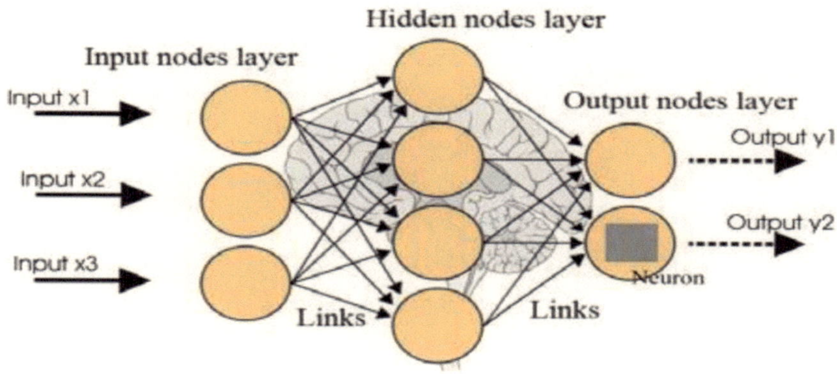

Figure 3.
Layers of ANN (artificial neural network) [23].

Human can perform	AI can perform
Speak and Listen	Speech recognition based on statistical learning system
Write and learn	Natural Language processing (NLP)
Eye can see	Computer vision or symbolic vision
Recognize the scene and create image	Image processing by symbolic learning
Understand the environment	Robotics
Ability to recognize pattern	Pattern recognition by Machine learning
Human brain formed by the networks of neurons	Artificial neural networks
Human memorize the past	Recurrent neural network (RNN) can use previous output as the input, so it remembers the data.
Recognize objects	Convolutional neural network (CNN) recognizes the object and also differentiates from others.

Table 1.
Similarity between human intelligence and artificial intelligence (AI).

and understands the environment through robotics. ML computes the large amount of data to get a solution to the problem in terms of pattern recognition. Statistical machine learning embedded with speech recognition and natural language processing. Deep learning recognizes objects by computer vision through convolution neural network (CNN) and memorize past by recurrent neural network (RNN). The schematic diagram of AI and its functions are shown in **Figure 4**.

The methods or techniques used for the AI are classifier and prediction. Classifier is an algorithm that implements classification; the classifiers are Perceptron, Naïve Bayes, Decision trees, Logistic regression, K nearest Neighbor, AANN/DL and support vector machine [24]. Perceptron is the basic building block of the neural network it breakdown the complex network to smaller and simpler pieces. The classifier used in the myoelectric prosthetic hand is LDA classifier, Quadratic discriminant classifier and Multilayer perceptron neural network with linear activation functions etc. LDA (linear discriminant classifier) is a simple one that helps to reduce the dimension of the algorithm for application of neural

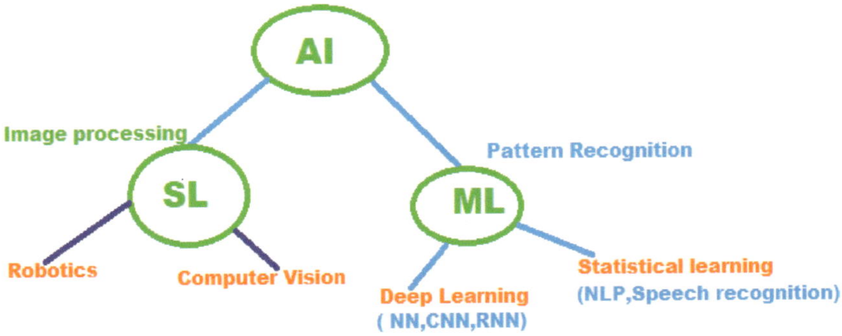

Figure 4.
AI and its functions.

network model. Prediction is a method to predict a pattern an output noise free data with a model from input data in hidden layer.

Examples: EMG CNN based prosthetic hand, EGG based Mind controlled prosthesis with sensory feedback, robotic arm, exoskeleton Orthosis.

4. Application of AI in prosthetics and orthotics

Implementation of artificial intelligence in controlling prostheses has increased drastically and thus enables the amputee to operate the prosthesis more desirably. Adaptive controlling would enable a system to perform closer to the desired output by adjusting the input with the help of a feedback system. Recently, a mind-controlled limb (type of myoelectric controlling) was introduced as the latest advancement in the artificial intelligence-aided control system. A joint project between the Pentagon and Johns Hopkins Applied Physics Laboratory (APL) has come up with a modular prosthetic limb which would be fully controlled by sensors implanted in the brain, and would even restore the sense of touch by sending electrical impulses from the limb back to the sensory cortex [25]. Chang et al. (2009) proposed a multilayer artificial neural network (ANN)-based model to discover the essential correlation between the intrinsic impaired neuromuscular activities of people with spina bifida (SB) and their extrinsic gait behaviors [26]. The application of AI in prosthetics and orthotics is divided into various subparts according to the involvement of the region that get affected i.e. Lower extremity prosthesis and Orthosis, Upper extremity Orthosis and prosthesis, and rehabilitation aids like motorized mobility devices.

4.1 AI in upper extremity prosthesis and orthosis

The artificial Intelligence in upper extremity prosthesis used as direct control and indirect control from the neural network by various signal, sensor, controller and algorithm. The control signals are coming from the human in the two form for operation of upper extremity prosthesis i.e. electromyography (EMG) and Electroencephalogram (EEG). Prior attempts at voluntary control of the elements of prosthesis have focused on the use of electromyography (EMG) signals from muscle groups that remain under voluntary control. Most of this work has centered on control systems for upper extremity prostheses. The first commercialized powered hand myoelectric prosthesis was introduced by USSR in 1960 [27]. The advancement in EMG control myoelectric prosthesis was with use of EMG pattern recognition based control strategy [28]. This approach allows the user to control the prosthesis with multiple degrees of freedom. The most advanced and developed neural machine interface technology was TMR or targeted muscle reinnervation [29].

The conventional Electromyography (EMG) technique uses bipolar surface electrodes, placed over the muscle belly of the targeted group of muscles. The electrodes are noninvasive, inexpensive, and readily incorporated into the socket of the prosthesis. These surface electrode have limitations like inability to record the signal from different muscle group at a time, inconsistency in signal magnitude and frequency, due to change in skin electrode interface associated in physiological and environmental modifications and also the EMG signals may encounter noise and interference from other tissues. Apart from these limitations it is easy to use by amputee and risk free. The amplitude of the EMG signal is mostly proportional to the contraction of the remaining muscle. To enhance the quality of the signal the Myoelectric control of prosthesis or other system utilizes the electrical action potential of the residual limb's muscles that are emitted during muscular contractions. These emissions are measurable on the skin surface at a microvolt level. The emissions are picked up by one or two electrodes and processed by band-pass filtering, rectifying, and low-pass filtering to get the envelope amplitude of EMG signal for use as control signals to the functional elements of the prosthesis. The myoelectric emissions are used only for control. In simultaneous control (muscle co contraction) and proportional control (fast and slow muscle contraction) controls the two different mode from wrist to terminal device and vice versa.

The advance method over the conventional technique of EMG signal which replace the complicated mode of switching is the pattern recognition. This new control approach is stranded on the assumption that an EMG pattern contains information about the proposed movements involved in a residual limb. Using a technique of pattern classification, a variety of different intended movements can be identified by distinguishing characteristics of EMG patterns. Once a pattern has been classified, the movement is implemented through the command sent to a prosthesis controller. EMG pattern-recognition-based prosthetic control method involves performing EMG measurement (to capture reliable and consistent myoelectric signals), feature extraction (to recollect the most important discriminating information from the EMG), classification (to predict one of a subset of intentional movements), and multifunctional prosthesis control (to implement the operation of prosthesis by the predicted class of movement) [30]. EMG pattern recognition block diagram of Trans radial prosthesis shown in **Figure 5**.

In pattern recognition control for a multifunctional prosthesis, multi-channel myoelectric recordings are needed to capture enough myoelectric pattern information. The number and placement of electrodes would mainly depend on how many classes of movements are demanded in a multi-functional prosthesis and how many residual muscles of an amputee are applicable for myoelectric control. For myoelectric transradial prostheses, the EMG signals are measured from residual muscles with a number of bipolar electrodes (8-16) which are placed on the circumference of the remaining forearm in which 8 of the 12 electrodes were uniformly placed around the proximal portion of the forearm and the other 4 electrodes were positioned on the distal end. A large circular electrode was placed on the elbow of the amputated arm as a ground [31].

For acquisition of EMG signal 50 Hz-60 Hz can be used to remove or reduce more low-frequency to increase the control stability of a multifunctional myoelectric prosthesis [32]. EMG feature extraction is performed on windowed EMG data, all EMG recordings channels are segmented into a series of analysis windows either with or without time overlap (WL (window length) is 100-250 ms) shown in **Figure 6** [33].

Overlapping analysis windows are used to maximally utilize the continuous stream of data and to produce a decision stream, for analysis, the duration of the overlapping (e.g., 50 ms) due to data buffering is the operational delay in real-time control and 50% of overlapping is suitable for the real time embedded system. The features are categorized as time domain (TD), frequency domain (FD) and time- frequency domain (TFD). The EMG features are extracted from each analysis window as a representation of EMG signal pattern. A feature set is extracted for each analysis window and all the recording channels, producing an L-dimensional feature vector. After computing the feature sets of all the channels, the entire EMG

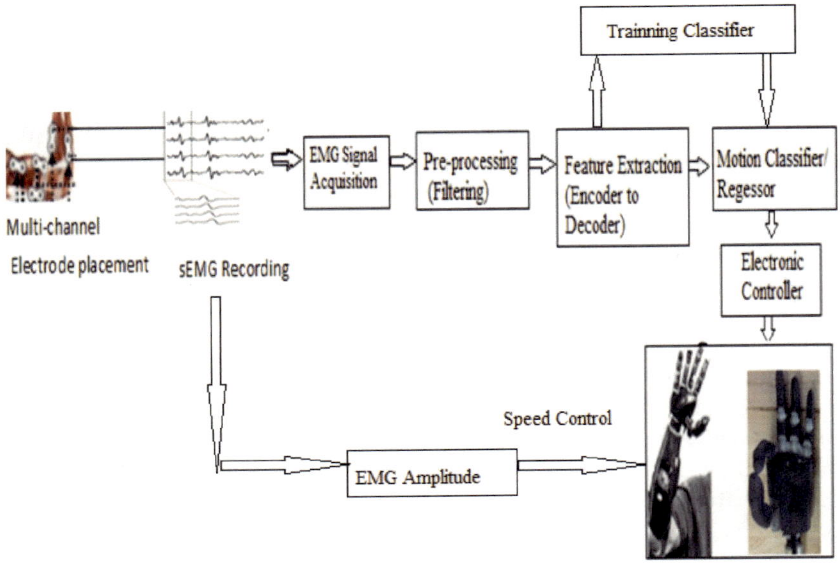

Figure 5.
Process of EMG pattern recognition control.

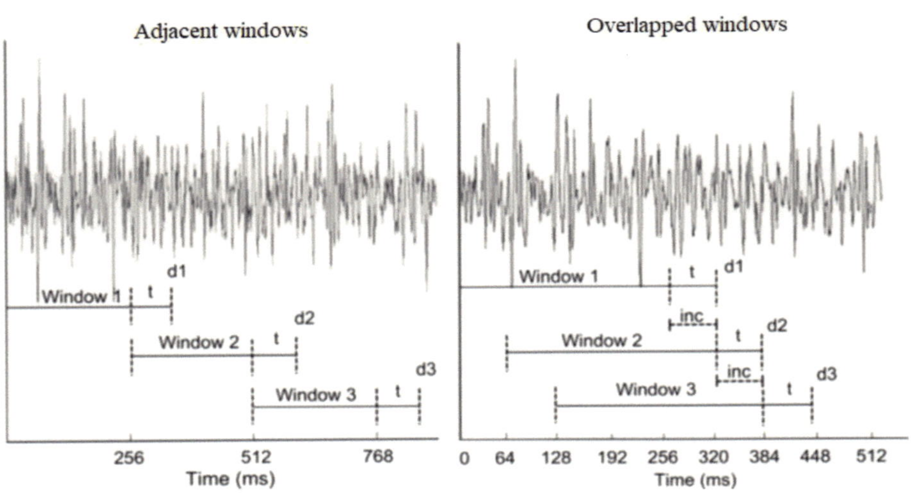

Figure 6.
Windowing techniques, time to process each window analysis is 't' and decisions (d1, d2, d3). In adjacent windows the processing time is less and the classifier is idle most of the time but in overlapping windows increase frequency of class decision because the analysis window slides with small increment (inc), the amount of overlap is equal to processing time which help the controller to process next class decision before the previous decission has been completed [33].

feature matrix (L × C × W, where *L*, *C*, and *W* are the number of features, the number of channels, and the number of analysis windows, respectively) from the training set is provided to a classifier for training shown in **Figure 7**. Example: The features extracted from four channels of surface EMG in each window is 44 and the data analyzed for the three windowed length, the EMG feature matrix for this situation (L × C × W = 44 × 4 × 3 i.e. L = 44, C = 4, W = 3).

The aim of pattern recognition based classifier is to discriminate the intended movements from the EMG recordings as accurately as possible. Many classification techniques have been investigated, including linear discriminate analysis, Bayesian statistical methods, artificial neural networks, and fuzzy logic [34, 35]. The LDA classifier is much simpler to implement and much faster to train without compromising the accuracy (>93%). Then the performance of a trained classifier in identifying a movement is evaluated using the testing data set and measured by the classification accuracy, which is defined as:

$$\frac{\text{Number correctly classified samples}}{\text{Total number of testing samples}} \times 100\% \qquad (4)$$

The classification accuracies in identifying all the classes of movements are averaged to calculate the overall classification accuracy for a subject uses convolutional neural network (CNN). Block diagram for classification and regression pattern shown in the **Figure 8** [36].

EMG pattern recognition based prosthesis control strategy is not suitable for people with shoulder disarticulation amputations because few muscles remain in their residual arm from which to extract myoelectric control signals. To address this challenge, a new neural machine interfacing (NMI) technology called targeted muscle reinnervation (TMR) have been proposed and developed at Rehabilitation Institute of Chicago (RIC), which has the ability to improve control performance of multifunctional myoelectric upper-limb prostheses shown in **Figure 9** [37].

TMR uses the remaining nerves from an amputated limb and transfers them onto substitute muscle groups that are not biomechanically functional because they are no longer attached to the missing arm. During this transfer procedure, target muscles are denervated so that they can be reinnervated by the residual arm nerves

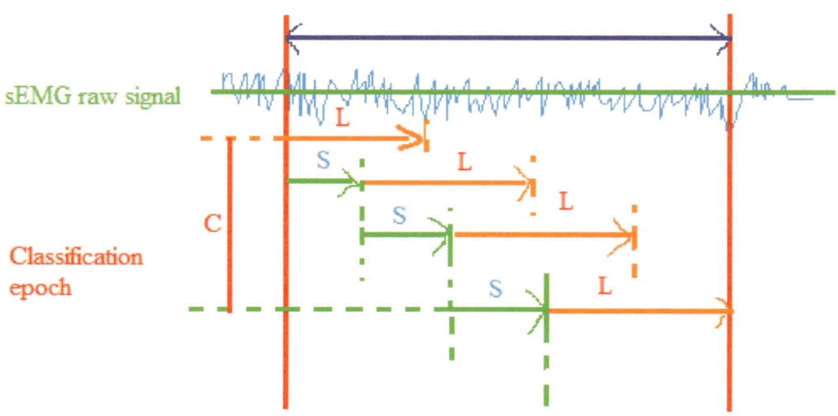

Figure 7.
EMG windowing in continuous feature extraction. Size of successive window for analysis is L, the sEMG data for classification is divided into C segments for every L that is the length of integrated samples as a feature extraction and the start point is shifted every S.

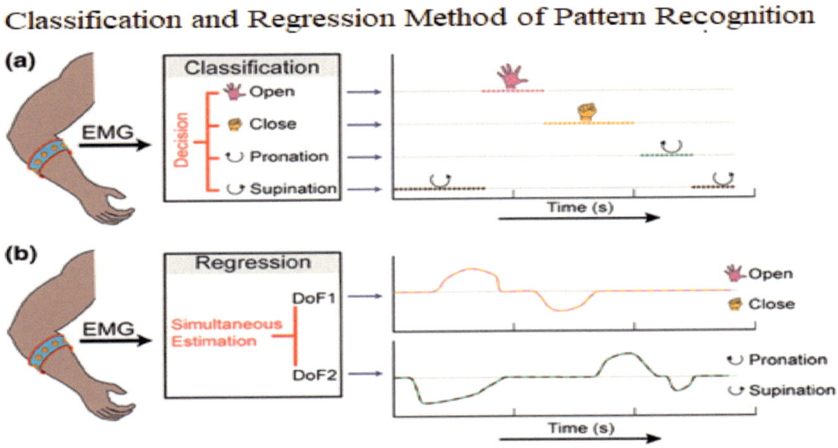

Figure 8.
a. Pattern recognition is able to classify different movement patterns, but only in sequence, which limits multifunctional control. b. Regression control is able to identify different movements at the same time, leading to more intuitive prosthetic control [36].

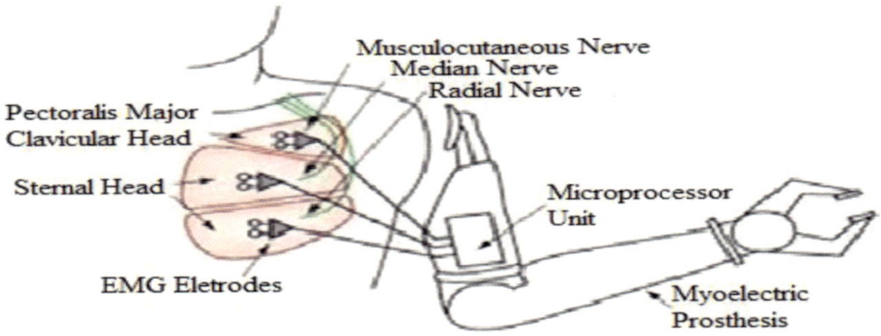

Figure 9.
Targeted muscle reinnervation (TMR) [37].

that previously traveled to the arm prior to amputation. The reinnervated muscles then assist as biological amplifiers of the amputated nerve motor commands. During the surgery subcutaneous tissue is removed that, surface EMG signals are optimized for power and focal recording.

Another advanced technique to control the multifunctional limb is Virtual reality (VR) based platforms have been developed for the purposes of development and performance quantification of multifunctional myoelectric prosthesis control system These VR platforms are designed to create an efficient, flexible, and user-friendly environment for prosthetic control algorithm development in the laboratory, application in a clinical setting, and eventual use in an embedded system. The major function modules of this platform include multi-electrode EMG recording (up to 16 channels), classifier training and testing in offline, virtual and physical prosthesis control in real time to regulate performance shown in **Figure 10** [38].

Apart from EMG signal the Electroencephalography (EEG) is the widely used non-invasive method by placing the electrode on the scalp for picking brain signal that has been utilized in brain machine interface (BCI/BMI) applications. It has high temporal resolution (about 1 ms) in comparison with other brainwave measurements such as electrocorticograms (ECoGs), magneto encephalograms

(MEGs), functional magnetic resonance imaging (fMRI) and near-infrared spectroscopy (fNIRS). The advanced prostheses may best control by EEG signal with BCI, connected by ANN. The neural signals associated with arm movements as control signals of artificial neuroprosthesis collected from either the cortex of brain directly or from residual nerves. The diagram of EEG based control and EMG pattern recognition based control in utilized in upper extremity prosthesis is shown in schematic **Figure 11**.

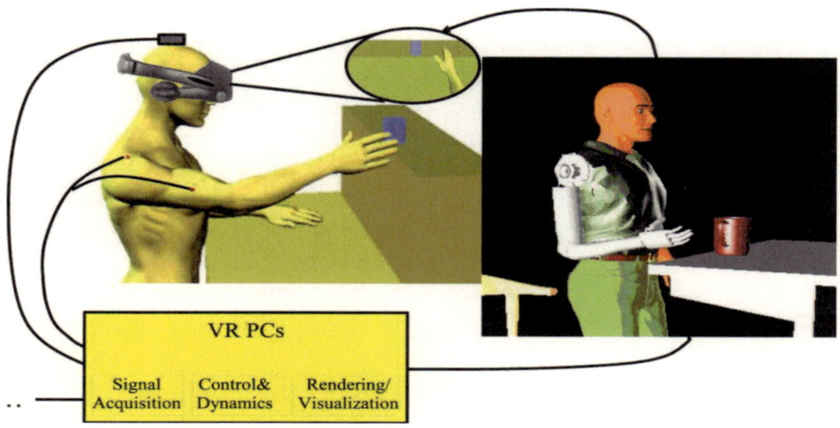

Figure 10.
Virtual reality system (VR), subjects can operate a simulated prosthetic arm to interact with virtual objects. Multiple input modalities such as motion tracking systems and EMG/EEG electrodes provide maximum flexibility when evaluating different control approaches. Figure shows a subject operating a prosthetic arm prototype in VR (right side). Subject controls the arm via real-time motion tracking (left side), and 3-D visual feedback is provided via stereoscopic goggles for closed loop operation [38].

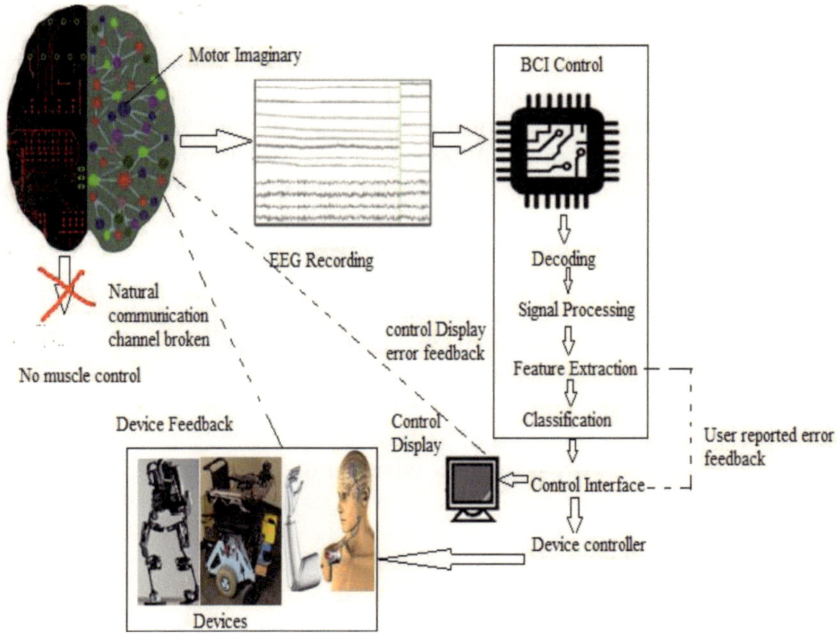

Figure 11.
Brain computer Interface (BCI), controlling prosthetic and orthotics devices.

Examples: Ottobock Dynamic Arm Plus is a combination of Myo Hand Vari Plus Speed terminal device and Wrist rotator with custom TMR socket which control the six DOF [39]. Mind or thought controlled prosthesis uses EEG signal and ANN.

4.1.1 Recent advancement in control strategy in upper extremity prosthetics and orthotics

Jafarzadeh M (2019) uses the novel deep convolutional neural network (6 convolutional layers and 2 deep layers) and FIFO memory for operation of prosthetic hand in real time. The novel CNN was implemented in Python 3.5 using tensor flow library [40].

Chih-Wei-Chen et al. (2009) developed BCI based hand Orthosis used cursor control interface with a simple LDA classifier, that classify the EEG signals to control the hand orthosis in to three state right, left and nil and the corresponding command as +1, −1 and 0. The four states of activities like grasp, open, holding and standby can control by these three commands. The +1 and − 1 command signifies grasp and open, command '0' is for standby mode depending on the feedback signals which are grasping force (F) and angular position (ϴ) collected from FSR and encoder [41].

4.2 AI in lower extremity prosthesis and orthosis

The first Artificial intelligence method used in the lower extremity as Intelligence prosthesis which is a knee joint that replace the hydraulic mechanism by combination of microprocessor controlled and hydraulic or pneumatic actuator.

The microprocessor as name suggests process the signal received by the first sensor known as knee angle sensor provides information about the knee's angle of flexion and extension and velocity of lateral and angular movement, unlike the human body, the sensor determines the direction of movement because of a magnetic implant and second sensor gathers information about weight placement.

Microprocessor receives the data or signals by the motion employed by the amputee and that data are analyzed and interpret to get the closer approximation to natural gait. This data provides information to the microprocessor about the device's position and the extent of its motion, which are essentially proprioceptive sensations. The data are stored in the memory of the microprocessor for the future use like a recurrent neural network (RNN). A series of wire networks which are similar in function to the body's nervous system. That is, it enables the sensors, microprocessor, servo motors, and hydraulic cylinder to communicate with each other. These networks connect the two sensors to the microprocessor, which transmits sensory data much like the ascending sensory pathways send information to the brain. The wires exiting the microprocessor leading to the servo motors carry "motion commands," mimicking the descending motor pathways which instruct muscles to contract and produce a desired movement.

As in the human nervous system, these wires are dedicated to specific communication circuits between the sensors, microprocessor, servo motors, and hydraulics. This computed data are used to control the resistance generated by the hydraulic cylinders through the small valve passes into and out of the cylinder which regulate extension and flexion of the knee joint in different sub phases of gait cycle. It controls knee joint motion from 0° to maximum 60-70^0. This mechanism helps the amputee to do various activities like stair climbing, jogging, running and walking in uneven terrain.

The microprocessor knee joint uses various algorithms to achieve gait symmetry, motion analysis, stumble control and comfort. These algorithm are control logic, Intent detection algorithm, Genetic algorithm, Fuzzy logic based classifier, Expectation maximization algorithm and Impedance control algorithm [42, 43]. The operation principle of a smart leg or intelligent prosthesis is shown in block diagram (**Figure 12**).

The Prosthetic knee joints uses this microprocessor control mechanism with machine learning Artificial Intelligence are Otto Bock's C leg (1997), OssurRheo knee (2005), Power knee by Ossur (2006), Self-learning knee by DAW Industries, Plie knee from freedom Innovation, Intelligent Prosthesis (IP) (Blatchford, United Kingdom), Linx (Endolite, Blatchford Inc. United Kingdom), Orion 2 (Endolite, Blatchford Inc. United Kingdom), X2 prostheses (Otto Bock Orthopedic Industry, Minneapolis, MN), X3 prostheses (Otto Bock Orthopedic Industry, Minneapolis, MN) etc.

The volitional EMG control robotics Transtibial prosthesis was developed in 2014 by Baojun Chen et al., which adapt the amputee to walk on slope with different angles. The combination of myoelectric and intrinsic controller reduces the fatigue of muscle and attention during walking [44]. The prototype design of prosthesis and schematic diagram of this mechanism showed in **Figure 13**.

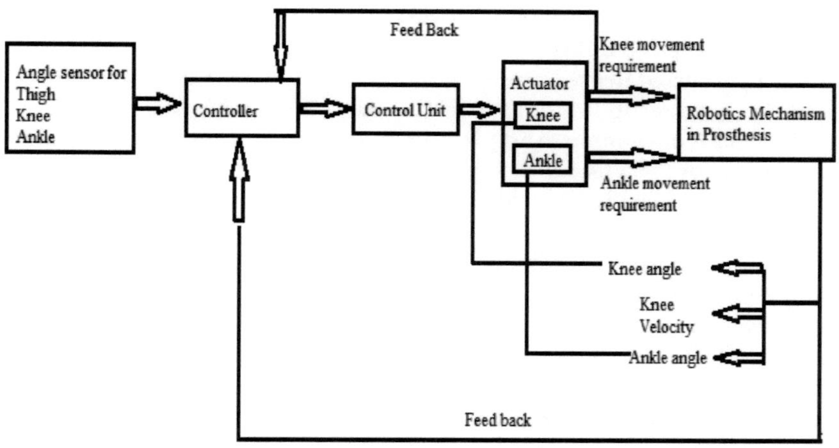

Figure 12.
Block diagram of controller based intelligent prosthesis.

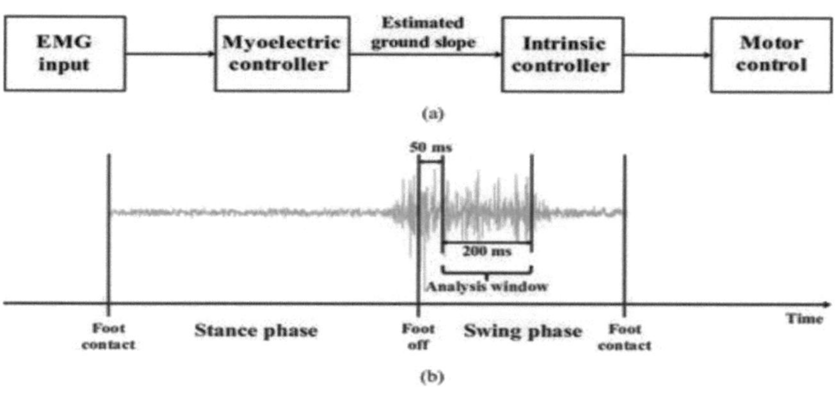

Figure 13.
(a) Schematic diagram of prosthesis control by integrating the proposed myoelectric controller with the intrinsic controller. (b) Strategy of extracting amputee users' movement intention with a 200-ms window in swing phase [44].

To mimic the normal foot and ankle motion several prosthetic feet uses AI mechanism are élan Foot (Blatchford, United Kingdom), iPED (developed by Martin Bionics LLC and licensed to College Park Industries), Proprio Foot (Össur, Iceland), Power Foot BiOM (developed at MIT and licensed to iWalk) and Meridium foot (Ottobock) etc. These feet are integrated with foot and ankle sensor to sense the terrain, angle and force required in different phases to mimic the normal foot.

Apart from EMG Control lower extremity prosthesis can be controlled by EEG signal using BCI, the example of EEG based control prosthesis is BiOM.

Lower Extremity Orthosis is a supportive device to the patients those have lost their function due to traumatic, neurologic and congenital abnormalities. The working principle of the Orthosis for the patient like hemiplegia, paraplegia and traumatic brain injury is changed vigorously with the implementation of artificial intelligence like functional electrical stimulation, Brain computer Interface and myoelectric controller. The concept of machine learning implemented in some sensor embedded stance control Orthosis which help the paraplegic to achieve near to normal gait with some limitations. The concept of functional electrical stimulation (FES) started in the year of 1960. This is used in case of damage of brain or spinal cord, stroke, Multiple Sclerosis (MS) and cerebral palsy.

The Functional electrical stimulation (FES) is the application of electrical stimulus to a paralyzed nerve or muscle to restore or achieve function. FES is most often used in neuro rehabilitation and is routinely paired with task-specific practice. Neuroprosthesis is a common example in orthotic substitution [45]. Control system can be open loop or Feed forward control, closed-loop or Feed backward control and adaptive control can be applied to both Feed forward and Feed backward controller. In open-loop controlled FES, the electrical stimulator controls the output and closed-loop FES employs joint or muscle position sensors to facilitate greater responsiveness to muscle fatigue, or to irregularities in the environment [46].

Electrodes act as interfaces between the electrical stimulator and the nervous system. The FES utilizes electrical current to stimulate muscle contraction so that the paralyzed muscles can start functioning again. The desired purpose is to stimulate a motor response (muscle contraction) through activation of a specific group of nerve fibers, typically using fibers of peripheral nerves. This may be achieved by the activation of motor efferent nerve fibers showed in **Figure 14**. FES uses Adaptive

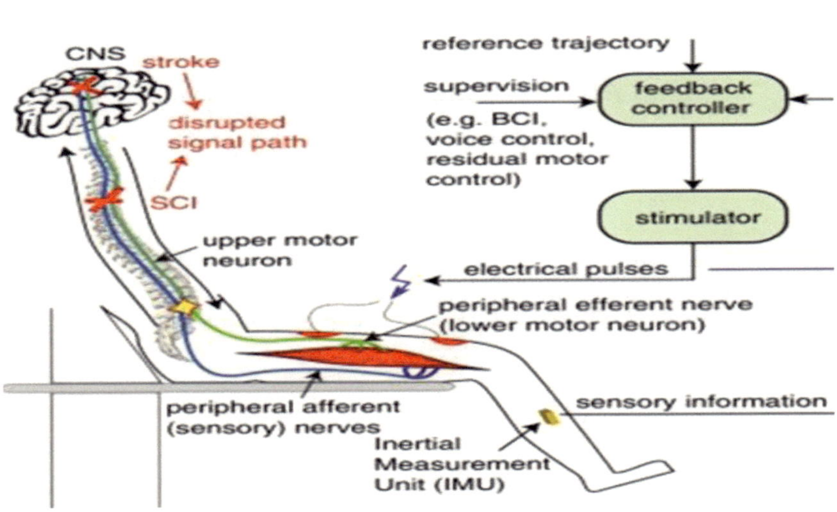

Figure 14.
Controlled functional electrical stimulation [47].

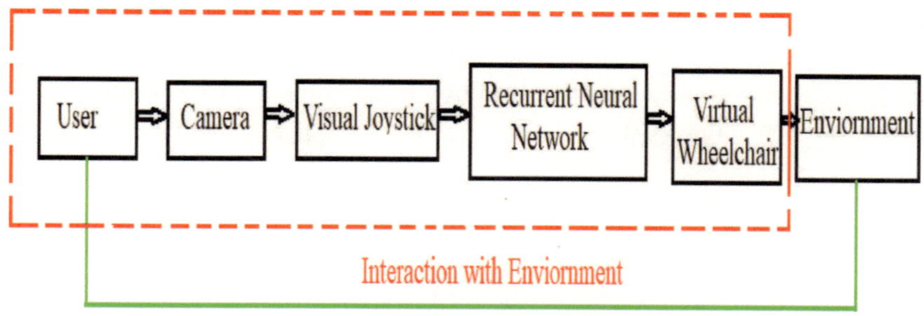

Figure 15.
Virtual simulation of visual joystick control wheelchair.

logic Network (ALN) and Inductive Learning Algorithm (IL) [47]. ALN is a type of artificial neural network for supervised learning which produces binary decision tree. This is a special type of feed forward multilayer perceptron the signal restricted to the Boolean logic. IL is a supervised learning produces decision tree in the form of IF, THEN, ELSE, etc. [48, 49].

AI implemented Gait Orthosis for spinal cord injury patients are powered ankle foot Orthosis (PAFO) and Exoskeletons. PAFO is incorporated with EMG controller to control the activity of soleus muscle to perform the actions of plantar flexion and inhibit the artificial dorsiflexion. Exoskeletons are uses BCI or EMG controller to control the orthotic devices [50].

4.3 AI in mobility devices

Wheel chair and walking aids is the important gadget for the disable to perform daily activities and transfer. In this robotic world the smart wheel chairs and intelligent walking aid reduced the area of work limitation. Application of artificial neural network in state of art robotics and AI technologies in smart wheels enhances the quality of life with ease in performance. The smart wheeler robotic wheelchair was developed by using Inverse Reinforcement Learning (IRL) techniques which was able to achieve maximum safety and set of tasks easily as compared to joystick control wheel chair [51]. Visual joystick control intelligent wheel chair is most advanced wheelchair prototype control by "Hand Gesture' incorporate recurrent neural network (RNN) in joystick control makes it a smart joystick having driving flexibility to different kind of disability [52]. The schematic diagram of virtual simulation for visual joystick control showed in **Figure 15**.

Smart cane is a boon for the visually impaired persons; it incorporated with raspberry PI 3 microcontroller, HC-SRC04 ultrasonic sensor for obstacle detection, WTV-SR IC recognition module for record and fix voice playback and GPS/GSM module to save different locations [53].

5. Conclusion

Human being is the most intelligent and complex engineered structure created by almighty. It is really a tough challenge for the Prosthetist & Orthotist to replicate its lost anatomical structure and function. However with advancement in the field of AI and robotics has created a ray of hope for millions of persons with disabilities. The application of AI in the field of prosthetics and orthotics are in the initial stage and not so widely being practiced. Many projects using AI are in prototype Stage and not yet commercialized. High costs of these devices are being major limitations as many Persons with disabilities cannot afford it. Government bodies, manufacturing unit and funding agencies must come forward and invest in this field so that the highest quality and latest technology must reach to larger population of disabled in an affordable cost.

Conflict of interest

The author does not have any conflict of interest.

Author details

Smita Nayak[1*] and Rajesh Kumar Das[2]

1 Pt. Deendayal Upadhyaya National Institute for Persons with Physical Disabilities (Divyangjan), New Delhi, India

2 Artificial Limbs Manufacturing Corporation of India, Kanpur, India

*Address all correspondence to: smitank7@gmail.com

References

[1] Jernigan TL, Stiles J. Construction of the human forebrain. Wiley Interdiscip Rev Cogn Sci. 2017;8(1-2):10.1002/wcs.409.

[2] Colombo M. Why build a virtual brain? Large-scale neural simulations as jump start for cognitive computing. Journal of Experimental & Theoretical Artificial Intelligence. 2017;29(2):361-70.

[3] Kandel ER, Schwartz JH, Jessell TM. Essentials of neural science and behavior. Norwalk, CT: Appleton & Lange; 1995.

[4] Rothwell J. From Neuron to Brain (Second Edition). A Cellular Approach to the Function of the Nervous System. J Neurol Neurosurg Psychiatry. 1985;48(2):196-7.

[5] Bezzi P, Volterra A. A neuron-glia signalling network in the active brain. Curr Opin Neurobiol. 2001;11(3):387-94.

[6] Demir MH, editor 1 Rehabilitation Technologies : Biomechatronics Point of View 2018.

[7] Akdogan E, Adli MA, Taçgin E, Bennett N, editors. A Human Machine Interface Design to Control an Intelligent Rehabilitation Robot System 2010.

[8] Taylor MB, Clark E, Offord EA, Baxter C. A comparison of energy expenditure by a high level trans-femoral amputee using the Intelligent Prosthesis and conventionally damped prosthetic limbs. Prosthet Orthot Int. 1996;20(2):116-21.

[9] Awad M, Dehghani-Sanij A, Moser D, Zahedi S. Inertia Properties of a Prosthetic Knee Mechanism 2015.

[10] Burner M, Program MUH. The C-Leg® and Proprioception: How a Microprocessor-controlled Knee Prosthesis Mimics the Human Body's System of Awareness of Position and Production of Movement 2011.

[11] Kyriazi NE, Pujol AO, Barcelona FdId, Universitat Politècnica de Catalunya. Departament d'Enginyeria de Sistemes AiII. AI and Prosthetics: Universitat Politècnica de Catalunya. Facultat d'Informàtica de Barcelona. Departament d'Enginyeria de Sistemes, Automàtica i Informàtica Industrial, 2016 (Master in Artificial Intelligence - MAI); 2016.

[12] Grabowski AM, D'Andrea SC, Herr HM, editors. BIONIC LEG PROSTHESIS EMULATES BIOLOGICAL ANKLE JOINT DURING WALKING 2011.

[13] Grabowski AM, D'Andrea SC. Effects of a powered ankle-foot prosthesis on kinetic loading of the unaffected leg during level-ground walking. Journal of NeuroEngineering and Rehabilitation. 2013;10:49 -

[14] Herr HM, Grabowski AM. Bionic ankle-foot prosthesis normalizes walking gait for persons with leg amputation. Proc Biol Sci. 2012;279(1728):457-64.

[15] Ernst M, Altenburg B, Bellmann M, Schmalz T. Standing on slopes - how current microprocessor-controlled prosthetic feet support transtibial and transfemoral amputees in an everyday task. J Neuroeng Rehabil. 2017;14(1):117.

[16] Clement RG, Bugler KE, Oliver CW. Bionic prosthetic hands: A review of present technology and future aspirations. Surgeon. 2011;9(6):336-40.

[17] Belter JT, Segil JL, Dollar AM, Weir RF. Mechanical design and performance specifications of anthropomorphic prosthetic hands: a review. Journal of rehabilitation research and development. 2013;50 5:599-618.

[18] Luchetti M, Cutti AG, Verni G, Sacchetti R, Rossi N. Impact of Michelangelo prosthetic hand: Findings from a crossover longitudinal study. J Rehabil Res Dev. 2015;52(5):605-18.

[19] Beyrouthy T, Kork SKA, Korbane JA, Abdulmonem A. EEG Mind controlled Smart Prosthetic Arm. 2016 IEEE International Conference on Emerging Technologies and Innovative Business Practices for the Transformation of Societies (EmergiTech). 2016:404-9.

[20] Sathya R, Abraham A. Comparison of Supervised and Unsupervised Learning Algorithms for Pattern Classification. International Journal of Advanced Research in Artificial Intelligence. 2013;2.

[21] Sidey-Gibbons JAM, Sidey-Gibbons CJ. Machine learning in medicine: a practical introduction. BMC Medical Research Methodology. 2019;19(1):64.

[22] Vasan G, Pilarski PM, editors. Learning from demonstration: Teaching a myoelectric prosthesis with an intact limb via reinforcement learning. 2017 International Conference on Rehabilitation Robotics (ICORR); 2017: IEEE.

[23] Spiers DL, editor Facial emotion detection using deep learning 2016.

[24] Sarker IH, Kayes ASM, Watters P. Effectiveness analysis of machine learning classification models for predicting personalized context-aware smartphone usage. Journal of Big Data. 2019;6(1):57.

[25] Bridges MM, Para MP, Mashner M, editors. Control System Architecture for the Modular Prosthetic Limb 2011.

[26] Yanbo H. Advances in Artificial Neural Networks – Methodological Development and Application. Algorithms. 2009;2.

[27] Kobrinski AE, Bolkhovitin SV, Voskoboinikova LM, Ioffe DM, Polyan EP, Popov BP, et al. Problems of bioelectric control. IFAC Proceedings Volumes. 1960;1(1):629-33.

[28] Momen K, Krishnan S, Chau T. Real-time classification of forearm electromyographic signals corresponding to user-selected intentional movements for multifunction prosthesis control. IEEE Trans Neural Syst Rehabil Eng. 2007;15(4):535-42.

[29] Li G. Electromyography Pattern-Recognition-Based Control of Powered Multifunctional Upper-Limb Prostheses. 2011.

[30] Yang Z, Chen Y. Surface EMG-based Sketching Recognition Using Two Analysis Windows and Gene Expression Programming. Front Neurosci. 2016;10:445.

[31] Yonghong H, Englehart KB, Hudgins B, Chan ADC. A Gaussian mixture model based classification scheme for myoelectric control of powered upper limb prostheses. IEEE Transactions on Biomedical Engineering. 2005;52(11):1801-11.

[32] Li G, Schultz AE, Kuiken TA. Quantifying pattern recognition-based myoelectric control of multifunctional transradial prostheses. IEEE Trans Neural Syst Rehabil Eng. 2010;18(2):185-92.

[33] Hakonen M, Piitulainen H, Visala A. Current state of digital signal processing in myoelectric interfaces and related applications. Biomedical Signal Processing and Control. 2015;18.

[34] Hargrove LJ, Englehart K, Hudgins B. A comparison of surface and intramuscular myoelectric signal classification. IEEE Trans Biomed Eng. 2007;54(5):847-53.

[35] Li G, Li Y, Yu L, Geng Y. Conditioning and Sampling Issues of EMG Signals in Motion Recognition of

Multifunctional Myoelectric Prostheses. Annals of Biomedical Engineering. 2011;39(6):1779-87.

[36] Roche AD, Rehbaum H, Farina D, Aszmann OC. Prosthetic Myoelectric Control Strategies: A Clinical Perspective. Current Surgery Reports. 2014;2(3):44.

[37] Ajiboye AB, Weir RF. A heuristic fuzzy logic approach to EMG pattern recognition for multifunctional prosthesis control. IEEE Trans Neural Syst Rehabil Eng. 2005;13(3):280-91.

[38] Hauschild M, Davoodi R, Loeb GE. A virtual reality environment for designing and fitting neural prosthetic limbs. IEEE Trans Neural Syst Rehabil Eng. 2007;15(1):9-15.

[39] Vujaklija I, Farina D, Aszmann O. New developments in prosthetic arm systems. Orthopedic Research and Reviews. 2016;20168:31-9.

[40] Jafarzadeh M, Hussey D, Tadesse Y. Deep learning approach to control of prosthetic hands with electromyography signals2019.

[41] Chen CW, Lin CC, Ming S. Hand Orthosis Controlled Using Brain-computer Interface. J of Med and Biol Eng. 2008;29:234-41.

[42] Stinus H. [Biomechanics and evaluation of the microprocessor-controlled C-Leg exoprosthesis knee joint]. Z Orthop Ihre Grenzgeb. 2000;138(3):278-82.

[43] Dedić R, Dindo H, editors. SmartLeg: An intelligent active robotic prosthesis for lower-limb amputees. 2011 XXIII International Symposium on Information, Communication and Automation Technologies; 2011 27-29 Oct. 2011.

[44] Chen B, Wang Q, Wang L. Adaptive Slope Walking With a Robotic Transtibial Prosthesis Based on Volitional EMG Control. IEEE/ASME Transactions on Mechatronics. 2014;20:1-12.

[45] Martin R, Sadowsky C, Obst K, Meyer B, McDonald J. Functional electrical stimulation in spinal cord injury:: from theory to practice. Top Spinal Cord Inj Rehabil. 2012;18(1):28-33.

[46] Bhatia D, Bansal G, Tewari R, Shukla KK. State of art: Functional Electrical Stimulation (FES). Int J of Biomedical Engineering and Technology. 2011;5:77-99.

[47] Schauer T. Sensing motion and muscle activity for feedback control of functional electrical stimulation: Ten years of experience in Berlin. Annual Reviews in Control. 2017.

[48] Martinez J. Neuroprostheses: Significance in Gait Rehabilitation. 2018. p. 427-46.

[49] Kostov A, Andrews B, Popović D, Stein R, Armstrong W. Machine learning in control of functional electrical stimulation systems for locomotion. IEEE transactions on bio-medical engineering. 1995;42:541-51.

[50] Rosen J, Fuchs MB, Arcan M. Performances of Hill-Type and Neural Network Muscle Models - Toward a Myosignal-Based Exoskeleton. Computers and biomedical research, an international journal. 1999;32 5:415-39.

[51] Pineau J, editor Designing Intelligent Wheelchairs: Reintegrating AI. AAAI Spring Symposium: Designing Intelligent Robots; 2013.

[52] Rabhi Y, Mrabet M, Fnaiech F. Intelligent Control Wheelchair Using a New Visual Joystick. Journal of Healthcare Engineering. 2018;2018:1-20.

[53] Ilag BN, Athave Y. A Design review of Smart Stick for the Blind Equipped with Obstacle Detection and Identification using Artificial Intelligence. International Journal of Computer Applications. 2019;182:55-60.

Chapter 3

Development of a Versatile Modular Platform for Aerial Manipulators

Nikolaos Evangeliou, Athanasios Tsoukalas,
Nikolaos Giakoumidis, Steffen Holter and Anthony Tzes

Abstract

The scope of this chapter is the development of an aerial manipulator platform using an octarotor drone with an attached manipulator. An on-board spherical camera provides visual information for the drone's surroundings, while a Pan-Tilt-Zoom camera system is used to track targets. A powerful computer with a GPU offers significant on-board computational power for the visual servoing of the aerial manipulator system. This vision system, along with the Inertial Management Unit based controller provides exemplary guidance in confined and outdoor spaces. Coupled with the manipulator's force sensing capabilities the system can interact with the environment. This aerial manipulation system is modular as far as attaching various payloads depending on the application (i.e., environmental sensing, facade cleaning and others, aerial netting for evader-drone geofencing, and others). Experimental studies using a motion capture system are offered to validate the system's efficiency.

Keywords: aerial manipulation, visual localization

1. Introduction

The introduction of drones has revolutionized many sectors, including but not limited to cinematography [1], search and rescue [2, 3], maintenance [4], surveillance [5, 6], delivery of goods and transportation [7, 8].

The main components of a drone are its Propelling System and its Flight Control Unit (FCU). The propelling system provides the necessary thrust to change the attitude of the drone, described by its pitch, roll and yaw angles, and thus its three dimensional motion. The dominant propelling system currently is composed by propellers driven by a brushless motor and an Electronic Speed Controller (ESC) combination. The FCU is the "brain" of the drone, since it issues the control commands to the ESCs for changing the attitude and the pose of a drone. It usually contains GPS receiver(s), accelerometer(s), gyroscope(s), magnetometer(s) and barometer(s) coupled to environment sensing devices like laser scanners to extract the current pose of the drone. The output of a FCU is computed by taking into account the current pose and the desired reference.

IntechOpen

Multi-rotor drones have been very popular among researchers with their naming typically by the rotor count (tricopters, quadcopters, hexacopters, and octacopters). The drone's thrust increases with the number of rotors allowing the lift of higher payloads at the expense of a reduced flight time, and power tethering systems are usually sought [9].

The majority of the off-the-self drones have a 1-2 kg payload capability with very few drones being capable of lifting an order of higher magnitude [10]. This is primarily due to the FCU's necessary tuning, the advanced ESCs and the need to abide to the laws imposed by each country's regulatory authority.

Pertaining to the described challenges, this chapter presents a drone that based on its mission can be modular in terms of software and hardware while lifting a high payload. The drone can operate either indoors or outdoors and has navigation and mapping capabilities as well as can interact with the environment through an attached robot manipulator.

In Section 2 the mechatronic design of the drone is presented, while in Section 3 the drone's software for localization is explained and evaluated. The drone's ability to perform either in a collaborating or an adversarial environment using computer vision is discussed in Section 4. The aerial manipulation concept is addressed in Section 5, followed by Concluding remarks.

2. Drone's mechatronic design

2.1 Drone electric power units

The developed octarotor drone has a take-off weight of 40 kg and a 30 min flight time. The drone's frame was designed and fabricated in collaboration with Vulcan UAV©. The authors' input on this aspect is related with both extending the bare-bone design of Vulcan to accommodate for payload carriage, as well as fabricating the final prototype and mounting all the additional modalities mentioned in the sequel. The backbone structure consists of three ø 25 mm, 1200 mm length aluminum tubes in a triangular cross sectional configuration. Four 575 mm length aluminum rectangular arms attached at each end of this structure and carry two motors in a coaxial configuration. The arms are fixed to the main frame using a 5 mm thick carbon plate. The resulting "H-frame" configuration can be visualized in **Figure 1**.

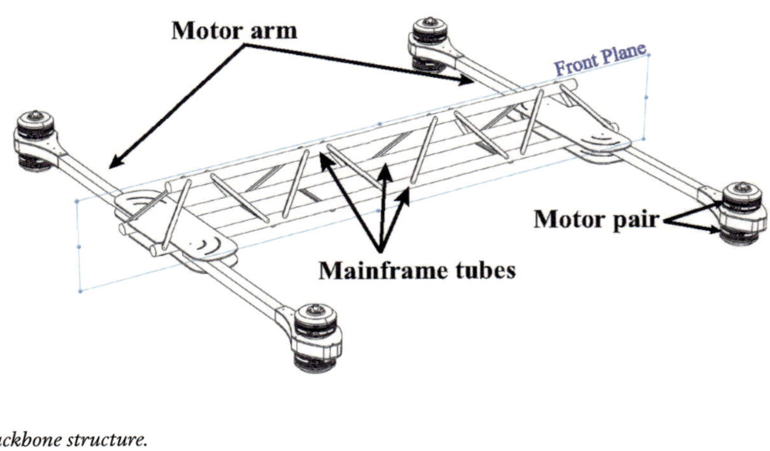

Figure 1.
Drone's backbone structure.

Although the lower motor provides 25% less thrust [11] it offers some redundancy against single motor failure. The selected 135 KV KDE© brushless motors coupled with ø71.12 cm custom designed carbon propellers, collectively provide 37.2 kg of thrust at 50% throttle input. The extra thrust can be used for rapid maneuvering of the drone and for exerting forces by the aerial manipulator shown in Section 5.

Power is provided by a 12S 22 Ah LiPo battery pair connected in parallel to the Power Distribution Board (PDB). At 50% thrust with full payload while hovering, the octarotor's motors sink 11.7 A each, resulting in a flight time of $\frac{2\times 22}{11.7\times 8}\times 60\,\text{min} = 26\,\text{min}$.

Two carbon rods of ø12 mm are fixed at the underside of the mainframe tubes for payload carriage. The maximum payload weight is 30 kg and can be easily dismantled from the main frame using quick release clamps. Similarly, the retractable landing gear assembly is attached with these clamps to the main frame tubes for enhanced modularity, as shown in **Figure 2**. The gear can retract within a $45^\circ - 80^\circ$ angle window using a Pulse Width Modulation (PWM) signal, provided by the FCU's rail pins, with a 50 Hz switching frequency. The landing gear operation is achieved via the MAVlink protocol command set [12].

Additional power for peripherals and sensing modalities can be supplied through a dedicated 750 W buck converter, mounted on the payload carrier assembly, as shown in **Figure 3**. The converter is contained within a custom 3D printed case and standard Unmanned Aerial Vehicle (UAV) $XT30, XT60$ connectors protrude to provide 24, 19 and 12 V respectively to the end-user.

2.2 Flight command unit and related software

The PixHawk Cube FCU was selected [13] featuring triple redundant dampened Inertial Measurement Units (IMUs), with a modular design and industrial standard I/O connectors. Additional telemetry and R/C circuits are deployed to enable monitoring and intervention and comply with flying regulations.

The *Here+* Global Navigation Satellite System (GNSS) [14] with Real-time kinematic (RTK) capabilities was selected for outdoor navigation and placed on top of a carbon fiber pole at a height of 35 cm from the main frame's top plane. For immunity to electromagnetic interference, the primary magnetometer of the flight controller is selected to be the build-in magnetometer module of the GNSS receiver.

A high processing power 8th generation Intel NUC i7-computing unit with 32 GB RAM and 1 TB SSD, shown in **Figure 3**, was mounted symmetrically to

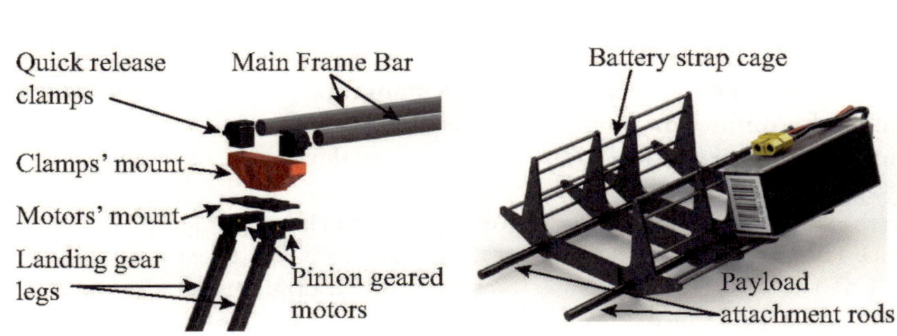

Figure 2.
Landing gear detail (left) and payload assembly with battery holder (right).

Figure 3.
Enhanced power distribution board (left) and i7-minicomputer (right).

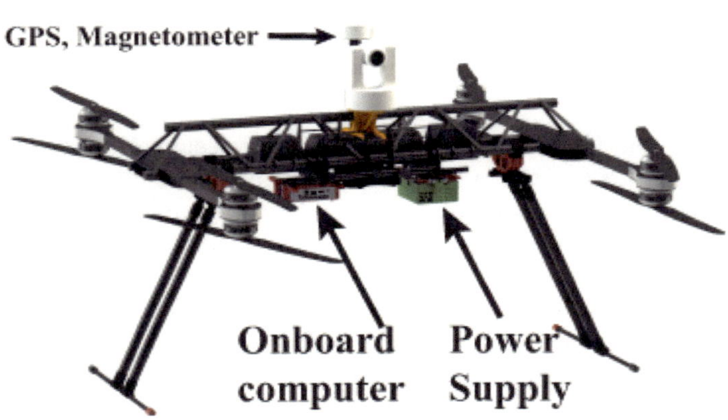

Figure 4.
Drone prototype.

the buck converter on the underside of the main frame. This 90 W computing unit allows for online computations on demanding tasks such as the visual object tracking methods of Section 4, as well as the easy development of autonomous flying applications.

On the software side, the ArduCopter flight stack [15] was selected to run on the FCU. The pose estimation is carried through a sophisticated Extended Kalman Filter (EKF) at 400 Hz. The Intel NUC companion computer is serially connected to the FCU at a baud rate of 1 Mbps and the communication packages are following the MAVlink protocol. The NUC's operating system was Ubuntu Linux 16.04 and all applications are developed through the Robot Operating System (ROS) and MAVROS [16] middleware with a 50 Hz refresh rate.

The developed drone without any payload can be visualized in **Figure 4**.

3. Drone localization

3.1 Drone outdoor localization using RTK GNSS

The RTK enhancement feature of GPS is used for outdoor localization purposes. This is due to the more precise positioning [17] because the of the GPS satellite measurements' correction using feedback from an additional stationary GPS module. The disadvantage of such systems is that their use is bounded to a significant pre-flight setup time which is inversely proportional to the achieved accuracy (cm range).

Although the internal loop of the flight controller operates at 400 Hz, the GPS receiver streams data at a lower rate of 5 Hz. In popular flight software such as ArduPilot, the aforementioned rate needs to be taken into consideration by the underlying EKFs running by the FCU. A typical comparison of the achieved accuracy using a drone in a hovering state can be seen in **Figure 5**.

The drone was flown in a hovering position with the RTK GPS module injecting measurements to the flight controller and the output of the FCU's EKF was compared with and without the presence of the injected RTK measurements. The red line represents the EKF's output based solely on the GPS signal, whilst the blue line indicates the same output when RTK correction (using a 30 min warmup period) is injected on the FCU.

The standard deviation was computed equal to 0.74 m, 0.47 m and 0.27 m for X, Y and Z respectively when no RTK correction was applied. Contrary to this, the same values with RTK injection were computed to equal 0.05 m, 0.02 m and 0.23 m respectively. It should be noted that there is no significant improvement in the Z-direction, indicating the need to use either a barometer or a laser sensor for ground clearance measurements.

3.2 Drone indoor localization

During indoor navigation: a) the lack of GPS guidance, b) pressure changes affecting the barometric sensor, and c) power lines affecting compass accuracy can severely affect the output of a FCU. With only the accelerometers and gyroscopes being unaffected, the injection of an external feedback source to the FCU is considered essential. Such feedback is usually based on visual techniques, such as those presented in [18, 19].

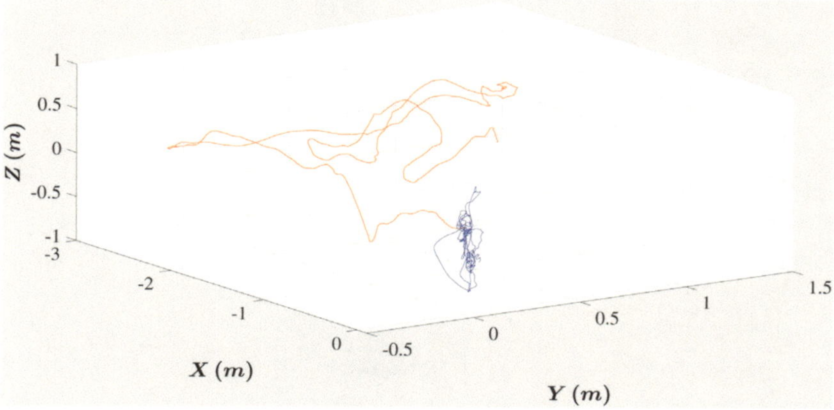

Figure 5.
Drone's EKF 3D-position output with (red) and without (blue) RTK correction.

For experimentation purposes, the used Motion Capture System (MoCaS) [20] injects measurements in the ArduCopter flight stack. The system comprises of 24 Vicon cameras uniformly scattered within an orthogonal space of $15 \times 5 \times 8 = (L \times W \times H)$ m. The utilized system allows simultaneous tracking of 100 objects at 120 Hz with sub-millimeter accuracy. Despite MoCaS's high refresh rate, the ArduCopter flight stack at the FCU accepts external positioning data at a 4 Hz streaming rate.

The utilized ROS software at the MoCaS operates at 25 Hz and can efficiently wirelessly stream the measurements to the drone's FCU. The latency time $t_d = t_V^C + t_C^{FCU} + t_{FCU}$, where $t_V^C \left(t_C^{FCU}\right)$ is the delay of data streaming to the companion computer (FCU), and t_{FCU} the delay of processing the data on the FCU. In the developed system typical measured values are $t_V^C \simeq 5\,ms$, $t_C^{FCU} = 20\,ms$ and $t_{FCU} = 40\,ms$, resulting in $t_d \simeq 75\,ms$.

Because of the MoCaS's efficiency, its weighing to the EKF is ten times larger compared to the GPS's weight when flying outdoors. Subsequently, the efficiency of the implementation is assessed by comparing EKF's position output with and without MoCaS's feedback injection. In **Figure 6** the drone's position error (in each axis) between the aforementioned two quantities is visualized, where the red, green and blue lines represent the error along the X Y and Z axes respectively.

Real time pose tracking is satisfactorily achieved and minor differences are attributed to the EKF's weighting of the accelerometer and gyroscope measurements during calculations.

4. Drone awareness of surrounding environment

An important parameter on aerial navigation is awareness of the surrounding environment including being in close proximity between cooperating or evasive drones [21, 22] to avoid potential contacts.

High accuracy awareness may not be feasible [23] and can become prohibitive in indoor environments; visual sensors along with Lidars can assist in this aspect. A spherical camera provides an all-around visualization of the surroundings and

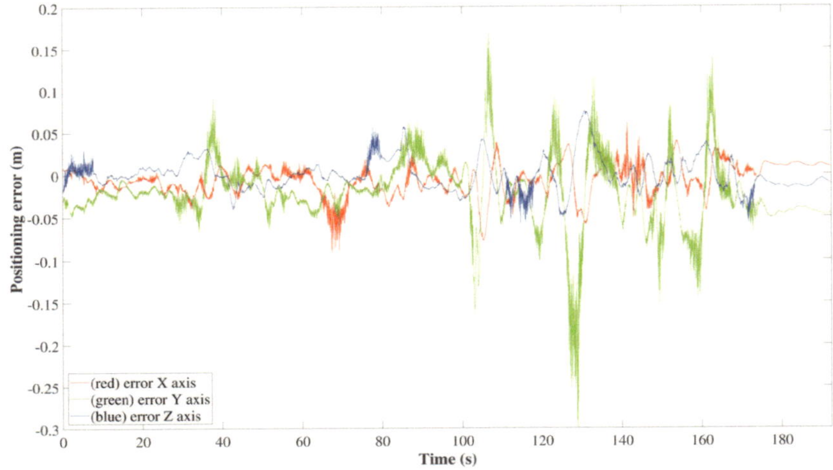

Figure 6.
Drone's EKF position error when MoCaS' measurements are not injected.

can detect neighboring targets. A Pan-Tilt-Zoom (PTZ) camera with a limited Field of View (FoV) can then provide a more accurate description of this target. The suggested target relies on the detection of moving objects. Correlation techniques and/or deep learning Visual Object Tracking (VOT) methods [24] are employed for this purpose.

4.1 Environment awareness using a spherical camera

Rather than using several cameras with a limited Field of View (FoV) to observe the surrounding space, a 360° FoV camera [25] is used. The spherical camera records images in a "spherical format" which is comprised of two wide-angle frames stitched together to form a virtual sphere [25]. The image can be rectified to the classic distortionless rectilinear format of a pinhole camera [26]. However, due to the nature of the "spherical format," it is preferable to split the image into smaller segments and rectify each one to achieve results closer to the pinhole camera model. Instead of splitting into equal sized square segments [27], each image is split into tiles based on orientation-independent circles. With every tile having a different a-priori known calibration, the rectification can be carried out for each one independently, without high computational cost. By applying the solution and rectifying the image in **Figure 7**, for a selection of $N = 12$ tiles, the resulting rectified partitions are visualized in **Figure 8**.

For the case of collaborating drones, it is assumed that each one carries passive markers for visual recognition. Subsequently, the rectified images are processed for identification of these markers [28–31] thus estimating the neighboring drone's pose. For improved pose extraction, the solution of a multi-marker rhombicuboctahedron formation arrangement [32] is assumed to be present in each target.

The experimental setup for evaluation consists of the spherical camera mounted in a 2.7 m protruding stick, which subsequently is mounted to the underside of the octarotor using the generic mount base discussed in Section 5. A rhombicuboctahedron arrangement with markers at its faces is attached to a DJI-Mavic drone. Both UAVs were located within the MoCaS test volume, as shown in **Figure 9**. The quadrotor drone was flown in a randomized trajectory near the vicinity of the octarotor.

In **Figure 10** the relative 3D-flight path between the drones is presented. The results recorded from the MoCaS and the visual ones are shown, where for the cases of detecting the marker the relative accuracy these measurements was 2.2 cm respectively.

Figure 7.
Spherical flat image.

Figure 8.
Rectified $N = 12$ *partitions for a single "spherical" frame.*

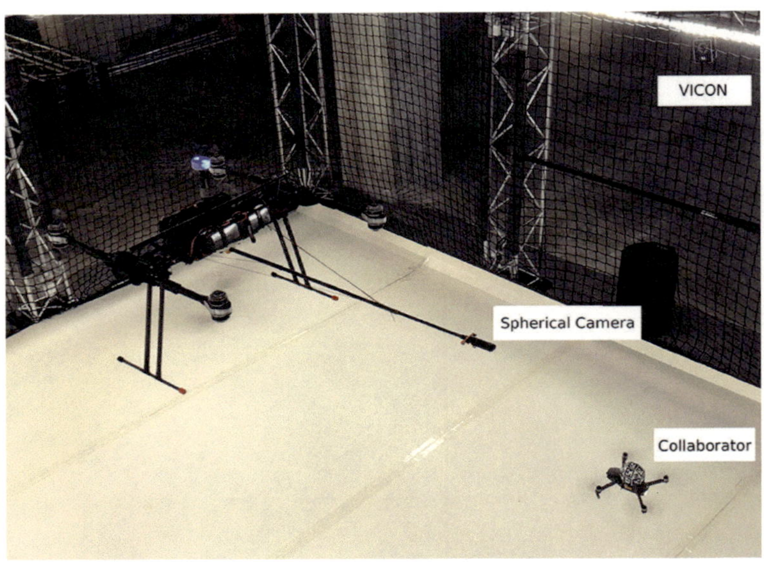

Figure 9.
Experimental setup for 360° camera relative localization.

4.2 Visual object tracking using a pan-tilt-zoom camera

Having identified the adversary or collaborating drone, a PTZ-camera is utilized to track its motion. This Visual Object Tracking (VOT) problem is challenging when the drone is occluded, thus Long Term Efficient (LTE) algorithms are sought for moving objects. Despite the development of Short Term Efficient (STE) algorithms [33] using

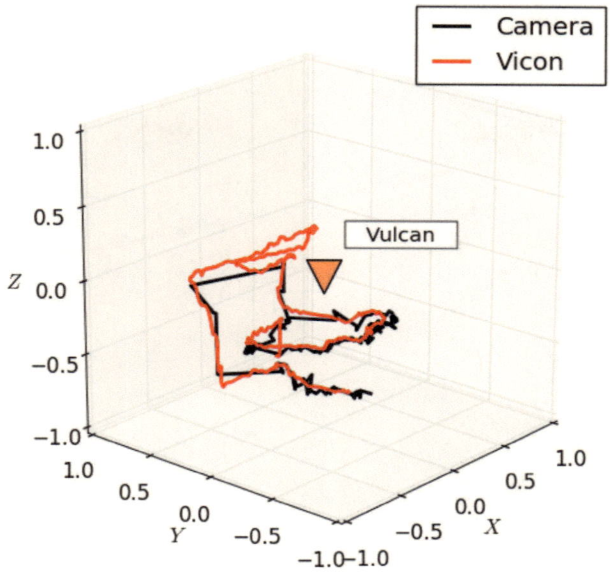

Figure 10.
3D-relative path inferred through the MoCaS and the visual method between two collaborating drones.

either correlation methods or deep learning ones, an initial bounding box containing the target is required. In the authors' case, the developed VOT algorithm employs two methods relying on a comparison: (a) between the tracking of the points transformed based on the PTZ-parameters and those using an optical flow, and (b) between the homography matrix transformed points and the optical flow.

The first method is based on the PTZ known motion and IMU's acceleration and gyroscope measurements (**Figure 11**), in order to estimate the motion of the pixels due to the motion of the camera in relation to the surroundings [34]. An IMU with triple accelerometers, gyroscopes, and magnetometers is attached to the PTZ-camera, as shown in **Figure 11**. While the enhancement provided by the PTZ camera allows for efficient VOT, the need to control its parameters (pan, tilt, and zoom) while placed on a floating base and at the presence of several occlusions needs to be addressed.

The objective is to provide the bounding box p_t of the approaching drone from the attached camera to the "Tracking drone" as shown in **Figure 12**. The IMU's sensors are sent to an embedded EKF to compute the camera's pan and tilt angles in the global coordinate system (and their angular velocities) at a 100 Hz rate. The angular velocities are used to compute the optical flow, and the angles are used for VOT purposes.

A GPU-based background subtraction technique eliminates the background pixels leaving only the moving object pixels. The bounding box encapsulates all pixels of the moving drone and the pan and tilt angles are adjusted to position the centroid of the moving bounding box at the image's center while the zoom is adjusted to enlarge this box. The communication between the i7-minicomputer and the PTZ-camera is shown in **Figure 13**, while the VOT algorithm is shown in **Figure 14**.

The feature points are recognized in each frame and the transformation matrix between successive frames follows along [35]; the formulas provide the transformation based on the PTZ-parameters and an augmentation is needed to account for the camera's translation, as provided by the on-board accelerometers. The pixels

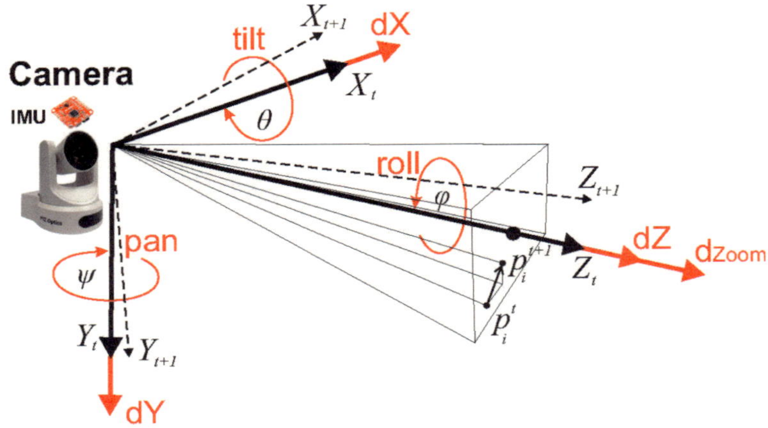

Figure 11.
PTZ-camera for visual object tracking.

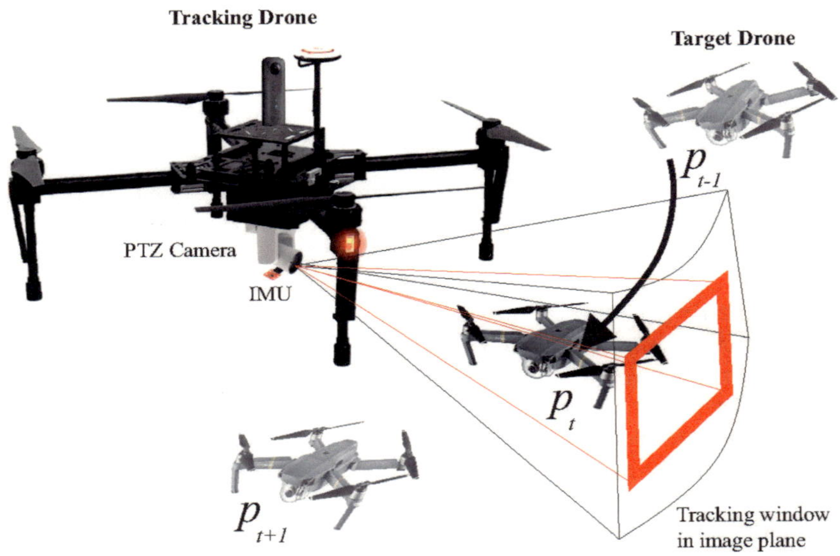

Figure 12.
Sample drone tracking setup.

that correspond to static background objects will follow the predicted motion by the camera motion and coincide to the positions predicted by an optical flow based estimation, while the rest will be classified as belonging to moving objects of interest (**Figure 15**). The computations for the optical flow parallels that of the Lucas-Kanade method using a pyramidal scheme with variable image resolutions [36]. The basic optical flow premise is to discover the positioning of an image feature in the previous frame, in the current frame captured by the camera.

The second method is relying only on visual feedback and homography calculations [37] between two successive frames and does not require either the PTZ or the IMU-measurements, as shown in **Figure 16**. Initially a set using "strong image features" is identified on the previous camera frame and an optical flow technique is used to estimate the position of the features in the current frame. The method involves the discovery of special image areas with specific characteristics.

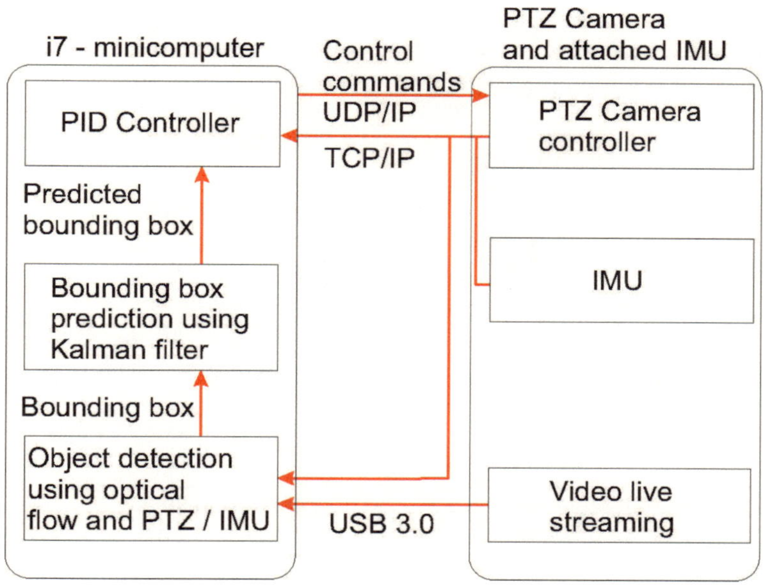

Figure 13.
PTZ-camera hardware tracking and control schematic.

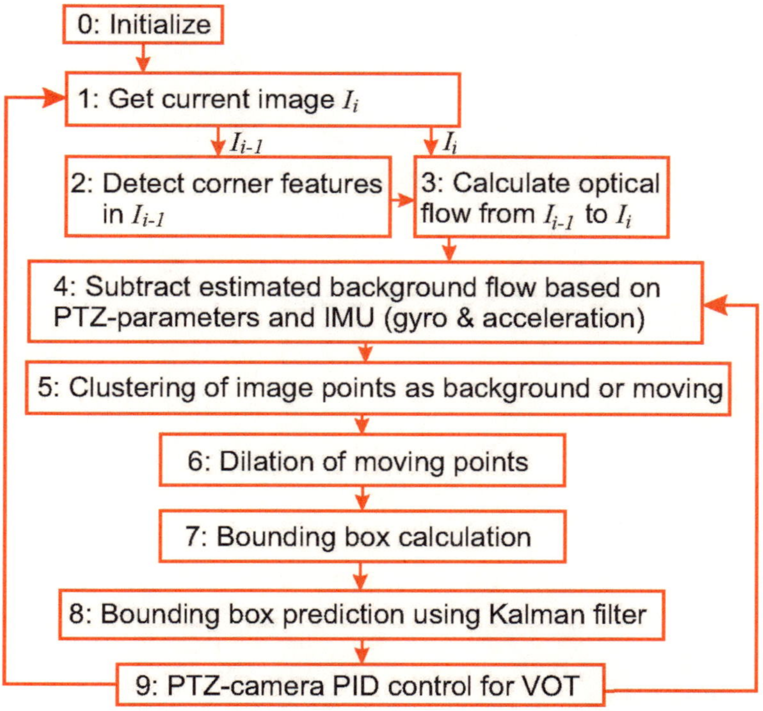

Figure 14.
Pan-tilt-zoom/IMU and optical flow VOT algorithm.

The algorithm used for finding the strong corners image features relies on the GPU-enhanced "goodFeaturestoTrack" [38]. Under the assumption that the background is formed by the majority of the pixels, a homography is calculated that transforms the features positions from the previous to the current frame; these correspond to the background pixels. The previous frame features are then transformed using the homography to get their position in the current frame. Herein, it

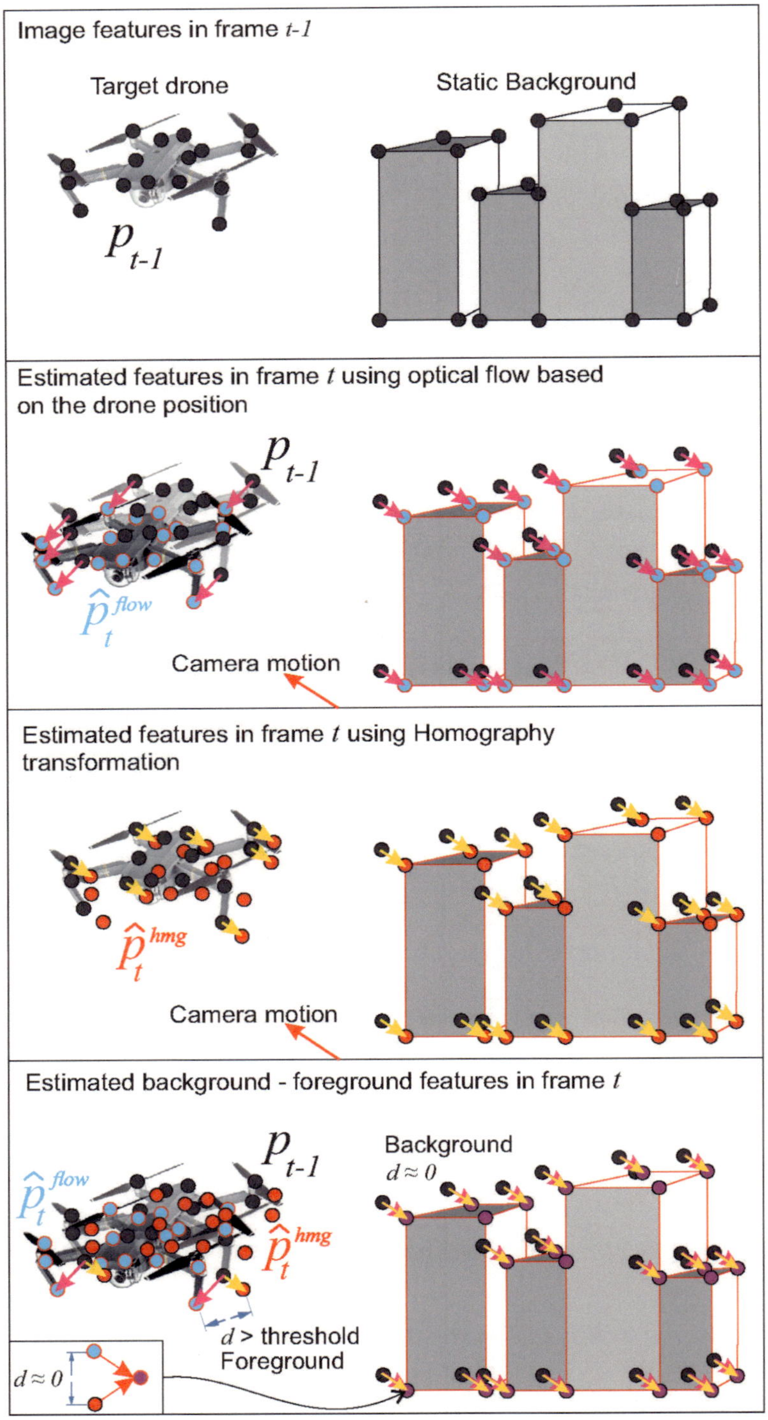

Figure 15.
Background/foreground estimation using Homography-based VOT.

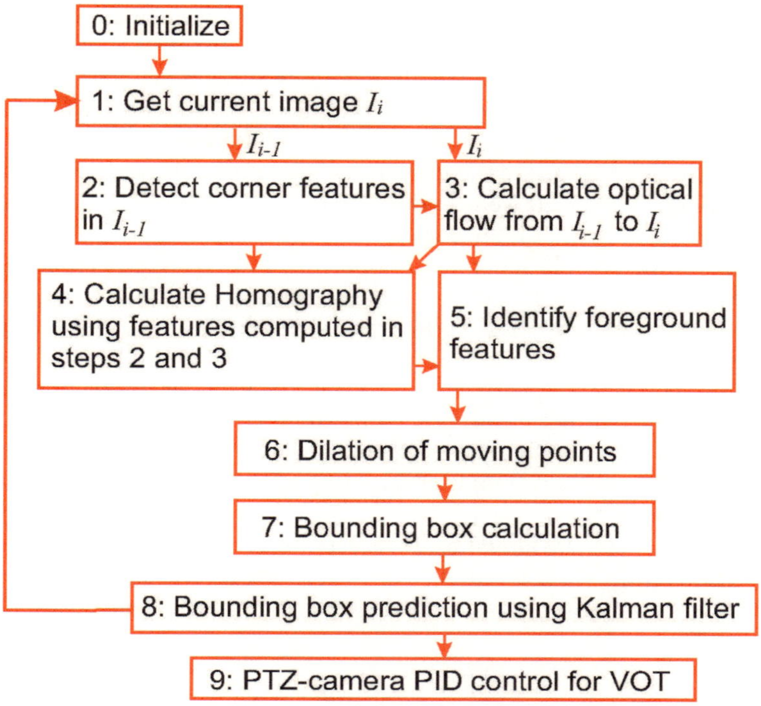

Figure 16.
Homography-based VOT.

is assumed that the background points transformed with the homography will coincide with the estimated ones by the optical flow, while the moving objects' features estimated by the optical flow will diverge from the homography transformed pixels.

One downside of the technique is that when the tracked object remains static and blends with the background it is unable to identify it. In this case, a fast correlation-based STE-tracker relying on the MOSSE algorithm [39], is also used in order to estimate the drone's position until new measurements of a moving drone are available. Several more robust but slower tracking algorithms were evaluated, including the KCF [40], CSRT [41], MIL [42], MedianFlow [43], TLD [44], and the MOSSE-algorithm was selected because of its fast implementation (600 Frames-per-Second (FpS)). A Kalman prediction scheme [45] was used to predict the bounding box and the one obtained from the MOSSE in the presence of noisy measurements of the moving object center, using a 2D-constant acceleration model for the estimated tracking window.

5. Aerial manipulation

A seven Degree-of -Freedom (DoF) robotic arm has been attached for exerting forces on surfaces in aerial manipulation tasks, such as grinding, cleaning or physical contact based inspection [46]. The Kinova Gen 2 Assistive 7DoF robot [47] was attached through a custom mount. This manipulator is characterized by a 2:1 weight to payload ratio, with the available payload at the end-effector being 1.2 kg grasped by the 3-finger gripper. Torque sensing is provided at each joint and these measurements along with the joint angles are communicated to the main computer at 100 Hz under ROS middleware.

For mounting the robot to the drone's payload attachment rods, a generic payload mount base was designed and manufactured. The base is firmly mounted to the drone's payload carrying rods utilizing quick attachment clamps. The construction material was selected to be T-6065 aluminum and features four 10 mm openings for attaching the payload. A second rigid base was similarly designed for attaching the robot's base to the generic payload mount base using 10 mm hex bolts. An exploded view of the entire mounting configuration can be visualized in **Figure 17**. The aerial manipulator is shown in **Figure 18**.

The indoor position hold scheme of Section 3.2 was expanded [48, 49] so as to utilize the manipulator in a surface ultra-sound scanning scenario. The surface is placed at 45° angle in an a-priori known position. After taking off the FCU retracts the landing gear (if commanded) and moves the manipulator to its joint angles $[180, 90, 180, -30, 90]^\circ$ respectively. On arrival to the desired setpoint pose, the manipulator's end tip comes into contact with the surface and the system hovers at the specified pose for some time for performing the area scan. The process is completed with the onboard computer initiating a landing after returning to the initial take-off position.

The described scheme is aimed for future use in the Abu Dhabi airport's Miedfiled Terminal [50], for scanning the integrity of critical structures such as facades and rooftop. **Figure 19** presents the hovering pose of the physical prototype

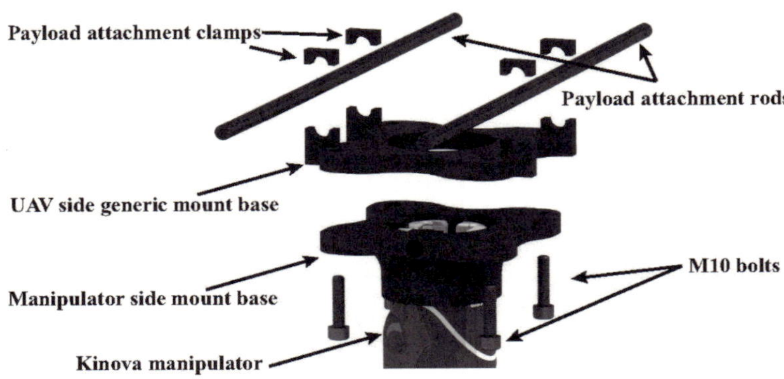

Figure 17.
Universal mount of robotic manipulators on aerial platform.

Figure 18.
Aerial manipulation system with PTZ-camera.

Figure 19.
Surface ultra-sound scanning utilizing aerial manipulation.

while scanning the surface, whilst the full video concept including moments of the experiment is available through the link given in [51].

6. Conclusions/discussion

In this chapter the mechatronic aspects (hardware and software) of a heavy lift drone are presented. This drone can operate either indoors or outdoors in an autonomous manner. Equipped with spherical and PTZ cameras, the drone has environment perception capabilities and can collaborate with other drones. A robot manipulator is attached at the drone for physical interaction purposes. The ability to carry out the aforementioned tasks in an accurate and modular manner depicts the efficiency of the system for future robotic aerial applications of increased complexity. However, many challenges are yet to be examined. The authors' aim is to focus future research on autonomous navigation in confined environments as well as high interaction forces aerial manipulation [52].

In aerial manipulation, the challenge lies with the forces at the tip of a stiff 7-DoF manipulator being directly transferred to the main UAS frame. Additionally, their orientation can be varying, depending on the pose of the manipulator. Thus, the ability of the aerial manipulator to robustly maintain its position and attitude while performing the task is mandatory. Compared to the depicted experimentation of this book chapter the induced forces from such operation are calculated to be in the area of 10 to 100 N. Subsequently, although the existing position controller of the ArduCopter flight stack is able to withhold a proper pose while ultrasound scanning of inclined areas, advanced control techniques [49] will be utilized in the sequel. The authors intend to test the efficiency of the built-in attitude controller of the ArduCopter flight stack, as well as exploit the adaptive backstepping control strategies in [48, 49] and other (model predictive) control techniques. The implementation of such controllers relies on the ability to directly control the angular velocity of the drone's motors independently, at rates greater or equal to 1 kHz.

Conflict of interest

The authors declare no conflict of interest.

Nomenclature

UAV	unmanned aerial vehicle
RC	remote control
PDB	power distribution board
IMU	inertial measurement unit
FCU	flight control unit
GNSS	global navigation satellite system
GPS	global positioning system
RTK	real-time kinematic
EKF	extended Kalman filter
ROS	robot operating system
DoF	degree-of-freedom
FoV	field-of-view
PTZ	pan-tilt-zoom
VOT	visual object tracking

Author details

Nikolaos Evangeliou[1*], Athanasios Tsoukalas[1], Nikolaos Giakoumidis[2], Steffen Holter[1] and Anthony Tzes[1]

1 Robotics and Intelligent Systems Control (RISC) Lab, Engineering Division, New York University Abu Dhabi, Abu Dhabi, 129188, United Arab Emirates

2 Kinesis Lab, Core Technology Platform, New York University Abu Dhabi, Abu Dhabi, 129188, United Arab Emirates

*Address all correspondence to: nikolaos.evangeliou@nyu.edu

References

[1] Mademlis I, Mygdalis V, Nikolaidis N, Pitas I. Challenges in Autonomous UAV Cinematography: An Overview. In: 2018 IEEE International Conference on Multimedia and Expo; 2018. p. 1-6.

[2] Nourbakhsh IR, Sycara K, Koes M, Yong M, Lewis M, Burion S. Human-robot teaming for search and rescue. IEEE Pervasive Computing. 2005;4(1):72-79.

[3] Hildmann H, Kovacs E. Using Unmanned Aerial Vehicles (UAVs) as Mobile Sensing Platforms (MSPs) for Disaster Response, Civil Security and Public Safety. Drones. 2019;3(3):59.

[4] Prada Delgado J, Ramon Soria P, Arrue BC, Ollero A. Bridge Mapping for Inspection Using an UAV Assisted by a Total Station. In: ROBOT 2017: Third Iberian Robotics Conference. Springer International Publishing; 2018. p. 309-319.

[5] Ding G, Wu Q, Zhang L, Lin Y, Tsiftsis TA, Yao Y. An Amateur Drone Surveillance System Based on the Cognitive Internet of Things. IEEE Communications Magazine. 2018;56(1):29-35.

[6] Ganesh Y, Raju R, Hegde R. Surveillance Drone for Landmine Detection. In: 2015 International Conference on Advanced Computing and Communications; 2015. p. 33-38.

[7] Barmpounakis E, Vlahogianni E, Golias J. Unmanned Aerial Aircraft Systems for transportation engineering: Current practice and future challenges. International Journal of Transportation Science and Technology. 2016;5(3): 111 – 122.

[8] Kellermann R, Biehle T, Fischer L. Drones for parcel and passenger transportation: A literature review. Transportation Research Interdisciplinary Perspectives. 2020;p. 100088.

[9] Papachristos C, Tzes A. The power-tethered UAV-UGV team: A collaborative strategy for navigation in partially-mapped environments. In: 22nd Mediterranean Conference on Control and Automation; 2014. p. 1153-1158.

[10] D130 X8 Titan Drone; 2020. Internet. Available from: https://www.foxtechfpv.com/d130-x8-titan.html.

[11] Otsuka H, Nagatani K. Thrust loss saving design of overlapping rotor arrangement on small multirotor unmanned aerial vehicles. In: 2016 IEEE International Conference on Robotics and Automation; 2016. p. 3242-3248.

[12] Koubâa A, Allouch A, Alajlan M, Javed Y, Belghith A, Khalgui M. Micro Air Vehicle Link (MAVlink) in a Nutshell: A Survey. IEEE Access. 2019;7:87658-87680.

[13] PixHawk 2 1 flight controller; 2020. Internet. Available from: http://www.proficnc.com/content/13-pixhawk2.

[14] Here+ RTK GNSS module; 2020. Internet. Available from: http://www.proficnc.com/content/12-here.

[15] The ArduCopter flight stack; 2020. Internet. Available from: https://ardupilot.org/copter/.

[16] MAVLink extendable communication node for ROS; 2020. Internet. Available from: http://wiki.ros.org/mavros.

[17] Rietdorf A, Daub C, Loef P. Precise positioning in real-time using navigation satellites and telecommunication. In: Proceedings of

The 3rd Workshop on Positioning and Communication; 2006.

[18] Chuang HM, He D, Namiki A. Autonomous Target Tracking of UAV Using High-Speed Visual Feedback. Applied Sciences. 2019;9(21):4552.

[19] Maravall D, de Lope J, Fuentes JP. Navigation and Self-Semantic Location of Drones in Indoor Environments by Combining the Visual Bug Algorithm and Entropy-Based Vision. Frontiers in Neurorobotics. 2017;11:46.

[20] Vicon Motion Tracking system; 2020. Internet. Available from: https://www.vicon.com/.

[21] Finn R, Wright D. Unmanned aircraft systems: Surveillance, ethics and privacy in civil applications. Computer Law & Security Review. 2012;28(2):184-194.

[22] Sappington RN, Acosta GA, Hassanalian M, Lee K, Morelli R. Drone stations in airports for runway and airplane inspection using image processing techniques. In: AIAA Aviation 2019 Forum; 2019. p. 3316.

[23] Woods AC, La HM. Dynamic target tracking and obstacle avoidance using a drone. In: International Symposium on Visual Computing. Springer; 2015. p. 857-866.

[24] Unlu HU, Niehaus PS, Chirita D, Evangeliou N, Tzes A. Deep Learning-based Visual Tracking of UAVs using a PTZ Camera System. In: IECON 2019 - 45th Annual Conference of the IEEE Industrial Electronics Society. vol. 1; 2019. p. 638-644.

[25] Aghayari S, Saadatseresht M, Omidalizarandi M, Neumann I. Geometric calibration of full spherical panoramic Ricoh-Theta camera. ISPRS Annals of the Photogrammetry, Remote Sensing and Spatial Information Sciences IV-1/W1 (2017). 2017;4:237-245.

[26] Young M. Pinhole optics. Applied Optics. 1971;10(12):2763-2767.

[27] Saff EB, Kuijlaars AB. Distributing many points on a sphere. The mathematical intelligencer. 1997;19(1):5-11.

[28] Munoz-Salinas R. Aruco: a minimal library for augmented reality applications based on OpenCV. Universidad de Córdoba. 2012;.

[29] Olson E. AprilTag: A robust and flexible visual fiducial system. In: 2011 IEEE International Conference on Robotics and Automation. IEEE; 2011. p. 3400-3407.

[30] Wang J, Olson E. AprilTag 2: Efficient and robust fiducial detection. In: 2016 IEEE/RSJ International Conference on Intelligent Robots and Systems. IEEE; 2016. p. 4193-4198.

[31] ALVAR A. Library for Virtual and Augmented Reality; 2016. Available from: http://virtual.vtt.fi/virtual/proj2/multimedia/alvar/.

[32] Tsoukalas A, Tzes A, Khorrami F. Relative Pose Estimation of Unmanned Aerial Systems. In: 2018 26th Mediterranean Conference on Control and Automation. IEEE; 2018. p. 155-160.

[33] Visual Object Tracking 2019; 2019. Internet. Available from: https://www.votchallenge.net/vot2019/trackers.html.

[34] Tsoukalas A, Evangeliou N, Giakoumidis N, Tzes A. Airborne visual tracking of UAVs with a Pan-Tilt-Zoom Camera. In: International Conference on Unmanned Aerial Systems; 2020. p. 66-71.

[35] Doyle DD, Jennings AL, Black JT. Optical flow background subtraction for real-time PTZ camera object

tracking. In: 2013 IEEE International Instrumentation and Measurement Technology Conference; 2013. p. 866-871.

[36] Bouguet JY. Pyramidal implementation of the Lucas Kanade feature tracker. Intel Corporation, Microprocessor Research Labs. 2000;.

[37] Szeliski R. Computer Vision: Algorithms and Applications. 1st ed. Berlin, Heidelberg: Springer-Verlag; 2010.

[38] Shi J, Tomasi C. Good Features to Track. In: Proceedings of IEEE Conference on Computer Vision and Pattern Recognition; 1994. p. 593-600.

[39] Bolme D, Beveridge J, Draper B, Lui Y. Visual object tracking using adaptive correlation filters. In: Proceedings of the IEEE Computer Society Conference on Computer Vision and Pattern Recognition; 2010. p. 2544-2550.

[40] Danelljan M, Khan FS, Felsberg M, v d Weijer J. Adaptive Color Attributes for Real-Time Visual Tracking. In: IEEE Conference on Computer Vision and Pattern Recognition; 2014. p. 1090-1097.

[41] Lukezic A, Vojir T, Cehovin L, Matas J, Kristan M. Discriminative Correlation Filter with Channel and Spatial Reliability. International Journal of Computer Vision. 2016 11;126.

[42] Babenko B, Yang M, Belongie S. Visual tracking with online Multiple Instance Learning. In: 2009 IEEE Conference on Computer Vision and Pattern Recognition; 2009. p. 983-990.

[43] Kalal Z, Mikolajczyk K, Matas J. Forward-Backward Error: Automatic Detection of Tracking Failures. In: 2010 20th International Conference on Pattern Recognition; 2010. p. 2756-2759.

[44] Kalal Z, Mikolajczyk K, Matas J. Tracking-Learning-Detection. IEEE Transactions on Pattern Analysis and Machine Intelligence. 2012 July;34(7):1409-1422.

[45] Paul Z, Musoff H. Fundamentals of Kalman Filtering: A Practical Approach. American Institute of Aeronautics and Astronautics, Inc.; 2015.

[46] Jung S, Song S, Youn P, Myung H. Multi-Layer Coverage Path Planner for Autonomous Structural Inspection of High-Rise Structures. In: IEEE/RSJ Int. Conference on Intelligent Robots and Systems; 2018. p. 1-9.

[47] Kinova Gen2 assistive manipulator; 2020. Internet. Available from: https://www.kinovarobotics.com/en/products/gen2-robot.

[48] Stergiopoulos Y, Kontouras E, Gkountas K, Giannousakis K, Tzes A. Modeling and control aspects of a UAV with an attached manipulator. In: 24th Mediterranean Conference on Control and Automation; 2016. p. 653-658.

[49] Chaikalis D, Khorrami F, Tzes A. Adaptive Control Approaches for an Unmanned Aerial Manipulation System. In: International Conference on Unmanned Aerial Systems; 2020. p. 498-503.

[50] UAE Miedfield Terminal Project; 2020. Internet. Available from: http://www.adac.ae/english/mtp/.

[51] UAE Miedfield Terminal structure maintenance promotional video; 2020. Internet. Available from: https://youtu.be/iD51n8OFUbg.

[52] Abbasi F, Mesbahi A, Velni JM. Coverage control of moving sensor networks with multiple regions of interest. In: American Control Conference; 2017. p. 3587-3592.

Chapter 4

Guidance-Based Motion Planning of Autonomous Systems

Bülent Özkan

Abstract

Motion planning is a significant stage in the control of autonomous systems. As an alternative method, guidance approach is proposed for the motion planning of those systems. In guided munitions, guidance laws determine the success of the guidance systems designed to steer systems such as missiles and guided bombs towards predefined targets. The guidance laws designated according to determinative agents such as the firing position of the munition, target type, and operational requirements try to provide the munition with arriving at the target point even under the disturbing effects. In this study, the applicability of the guidance laws to autonomous systems is investigated in a manner similar to the approach for the guided munitions. For this purpose, the motion planning of the selected robotic arm, tracked land vehicle, and quadrotor is tried to be performed in order to move them to predefined target points. Having designed the control systems compatible to the selected guidance laws for the considered systems, the corresponding guidance scheme is constructed. Eventually, after conducting the relevant computer simulations, it is observed that the desired target chase can be made in a successive manner for all cases.

Keywords: motion planning, guidance, guidance law, linear homing, autonomous system

1. Introduction

Motion planning constitutes one of the primary stages in the control of autonomous systems. The autonomous systems can perform their planned tasks under several environmental conditions as per the designated motion planning algorithms in accordance certain performance criteria. The mentioned criteria may include minimum energy or minimum time consumption and shortest path length. For the motion planning purpose, several different algorithms are proposed by relevant researchers. These methods have certain advantages and disadvantages over the others [1–4].

As an alternative approach, guidance schemes can be used in motion planning. Those schemes involve an upper-level guidance algorithm and a lower-level control system. In fact, guidance and control loops can be introduced as officer and soldier, respectively. In other words, as the guidance algorithm behaves as the "master," the control system takes the "slave" role in this scene [5].

The guidance schemes are widely implemented to munitions. Guidance and control systems are designed in a compatible manner with munition dynamics so that the munitions including missiles and guided bombs can carry payloads towards specified target points as planned. The guidance part of the mentioned guidance and control system constitutes the kinematic relationships established as per the relative position between the munition and intended target point, while the control system is the closed-loop control system constructed based on the dynamic model of the munition under consideration in order to realize the guidance commands generated by the guidance part. In this extent, the guidance approach enrolls as a motion planning scheme for the munition [5].

The type of the command yielded by the guidance system depends on the selected guidance law. Namely, as the output of the proportional navigation guidance law which constitutes the most widely used guidance law in guided munitions is the lateral component of the linear acceleration vector of the munition or change of the lateral angular component of the linear velocity vector in time as per the application, the command of the body pursuit guidance law becomes the components of the angle between the body longitudinal axis of the munition and the lateral axis of the Earth-fixed frame [5, 6].

Guidance laws designated to move the guided munitions towards specified target points make their motion planning in a sense. Regarding this property, there seems no serious obstacle on their implementation on autonomous systems other than guided munition. In this study, the orientation of the sample autonomous mechatronic systems involving a robotic arm, tracked land vehicle, and quadrotor to predefined target points using the linear homing guidance (LHG) law and the relevant computer simulations is carried out. Here, these systems are chosen as very common systems encountered in the physical world. In the considered cases, the LHG law is an angle-based approach, and it takes the flight path angle components of the systems into consideration. Also, the selection of the mentioned mechatronic systems allows evaluating the convenience of the proposed approach in the planar and spatial engagement situations. Moreover, single- and two-stage control systems are utilized in accordance with the LHG law.

Guidance-based motion planning schemes are developed for certain robotic arm configurations [7]. In this scene, the indicated strategy allows the operators to run the moving belt of the robotic arm-belt assembly line within a mounting line in a more continuous and faster manner than the usual methods. Moreover, it may become to suppress or at least minimize the drawbacks of the conventional approaches by regarding the guidance-based motion planning method. Namely, although many conventional methods require the belt assembly to halt at intermediate placing instants, the guidance-based approach makes possible to place the objects under consideration onto the belt while it remains running during operation [8].

Different from wheeled vehicles, tracked land vehicles are directed as per a sliding motion which depends on the rotation of the vehicle about an instantaneous rotation center. In other words, they can be oriented to left or right by rotating about their instantaneous rotation center in the convenient sense. Regarding the motion of these vehicles on soil surfaces especially, the motion planning becomes harder. As a remedy to this inconvenience, the guidance-based path planning approach is proposed in the present study [8].

As the third application, the motion planning of a quadrotor which is intended to carry a payload from a stationary initial point to a prescribed moving land platform at a moderate distance for either military or civil purpose is investigated. Here, the payload can be munition, food, or first aid material. Since the moving platform specified as the target is assumed to be far away from the initial point, it is desired for the quadrotor to catch it within the shortest time duration possible and consume the energy at a minimum level [8, 9].

In the computer simulations in which the planar motion of the robotic arm and tracked land vehicle and the spatial motion of the quadrotor are taken into consideration, it is assumed that the targets are moving along specified trajectories. Here, regarding the motion characteristics of the autonomous systems dealt with, the robotic arm operates on the horizontal plane, and the vertical displacement of the tracked land vehicle is in a negligible level compared to its longitudinal and lateral motion on the ground plane. Thus, the motion profiles are described on a plane for both the systems. Unlike them, the quadrotor flies in the sky towards all three directions. This fact leads to handle its dynamic behavior in a three-dimensional space. Having completed the computer simulations, it is observed that all three systems can be carried to the intended target points by LHG law [8].

As a motion planning strategy, guidance approach can be applied to service robots which are utilized to accomplish certain motion profiles apart from the industrial systems. In this way, it is intended to perform hazardous, tedious, and time-consuming tasks in a more efficient and accurate manner in daily use. The mentioned category for service robots, actually, involves not only articulated robot manipulators but also moving and flying autonomous structures as well.

2. Description of the systems

2.1 Description of the robotic arm

The schematic views of the robotic arm and mounting line, i.e., trajectory, containing the slot on which the component grasped by the gripper of the arm is placed is given in **Figure 1**.

The system whose schematic view is presented in **Figure 1** consists of a two-degree-of-freedom robotic arm and moving mounting line. Here, the object to be put on the slot on the mounting line by the gripper of the robotic arm is taken in spherical geometry, and thus its orientation is ignored. This way, the degree of freedom of the carried object is reduced to two. In other words, it becomes possible to define the instantaneous planar location of the object by regarding the lateral and vertical position components of a point, i.e., point P, on the object. So, a robotic arm is required with minimum degree of freedom of two in order for the object with two degrees of freedom to be carried upon a specified point on the plane without any control deficiency. The definitions made in **Figure 1** are listed as follows [8]:

x and y: lateral and vertical axes of the Earth-fixed frame symbolized by F_O.

$\vec{u}_1^{(0)}$ and $\vec{u}_2^{(0)}$: unit vectors denoting the x- and y-axis of F_O.

O and A: joints of the robotic arm.

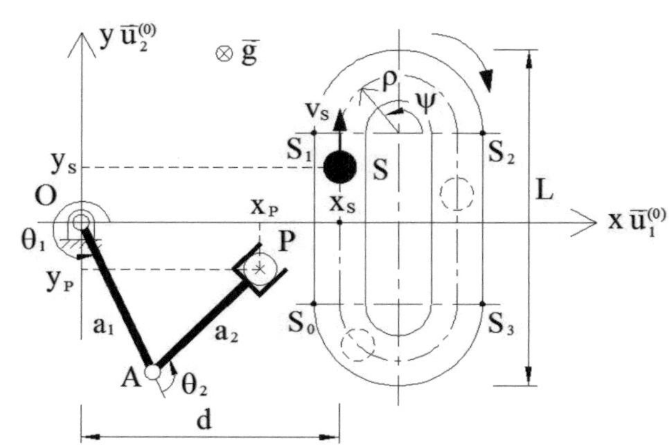

Figure 1.
System consisting of a robotic arm and moving mounting line [8].

a_1 and a_2: lengths of the first and second links of the arm.
θ_1 and θ_2: first and second joint angles of the robotic arm.
P: point defined on the gripper of the robotic arm.
x_P and y_P: lateral and vertical position components of point P.
S: midpoint of the slot on the mounting platform.
S_i: form changing points of the mounting line (i = 1, 2, 3, and 4).
v_S: speed of the slot on the mounting line.
x_S and y_S: lateral and vertical position components of point S.
ρ: turn radius of the mounting line.
ψ: rotation angle on the circular tip parts of the mounting line.
L: total length of the mounting line.
d: perpendicular distance between the connection point of the robotic arm to the ground and the point of the mounting line closest to that connection point.
$\vec{g}$: gravity vector (g = 9.81 m/s^2).

Regarding these definitions, the mathematical model of the robotic arm can be expressed in a compact matrix form as follows [8]:

$$\overline{T} = \hat{M}(\overline{\theta})\ddot{\overline{\theta}} + \hat{H}\left(\dot{\overline{\theta}},\overline{\theta}\right)\dot{\overline{\theta}} \tag{1}$$

where, as $\overline{\theta} = [\theta_1 \;\; \theta_2]^T$ and $\overline{T} = [T_1 \quad T_2]^T$, $\hat{M}(\overline{\theta})$ and $\hat{H}\left(\dot{\overline{\theta}},\overline{\theta}\right)$ which denote the inertia matrix and compound friction and Coriolis effect matrix, respectively, are defined as $\hat{M}(\overline{\theta}) = \begin{bmatrix} m_{11} & m_{12} \\ m_{12} & m_{22} \end{bmatrix}$ and $\hat{H}\left(\dot{\overline{\theta}},\overline{\theta}\right) = \begin{bmatrix} h_{11} & h_{12} \\ h_{21} & h_{22} \end{bmatrix}$ with $m_{11} = m_1 d_1^2 + m_2\left[a_1^2 + d_2^2 + 2a_1 d_2 \cos(\theta_2)\right] + I_{c1} + I_{c2}$, $m_{12} = m_2 d_2\left[d_2 + a_1 \cos(\theta_2)\right] + I_{c2}$, $m_{22} = m_2 d_2^2 + I_{c2}$, $h_{11} = b_1 - 2m_2 a_1 d_2 \dot{\theta}_2 \sin(\theta_2)$, $h_{12} = b_2 - m_2 a_1 d_2 \dot{\theta}_2 \sin(\theta_2)$, $h_{21} = m_2 a_1 d_2 \dot{\theta}_1 \sin(\theta_2)$, and $h_{22} = b_2$ [8].

In the shorthand definitions above, m_1, m_2, Ic_1, and Ic_2 denote the masses of the first and second links of the manipulator and the moments of inertia of these links with respect to their mass centers indicated by C_1 and C_2, respectively. Also, b_1 and b_2 represent the viscous friction coefficients at the first and second joints as well as the definitions of $d_1 = |OC_1|$ and $d_2 = |AC_2|$ [8].

2.2 Description of the tracked land vehicle

As the second system examined, the schematic description of the tracked land vehicle is shown in **Figure 2**. The explanations of the quantities labeled in **Figure 2** are given as follows [8]:

O, G, and C: origin of F_O, mass canter of the tracked vehicle, and instantaneous rotation center of the vehicle.
$\vec{u}_i^{(0)}$: unit vectors of F_O.
$\vec{u}_i^{(b)}$: unit vectors of the tracked land vehicle frame, i.e., F_b.
x and y: position components of point G on F_O.
ψ: orientation angle of the vehicle on the vertical plane.
$\vec{r}_{G/O}$: relative position of point G with respect to point O.
$\vec{r}_{G/C}$: relative position of point G with respect to point C.
x_C and y_C: position components point C in F_b.
a, b, c, d, and v: dimensional parameters of the vehicle.
m: mass of the tracked land vehicle.
$\vec{W}$: weight vector of the tracked land vehicle.

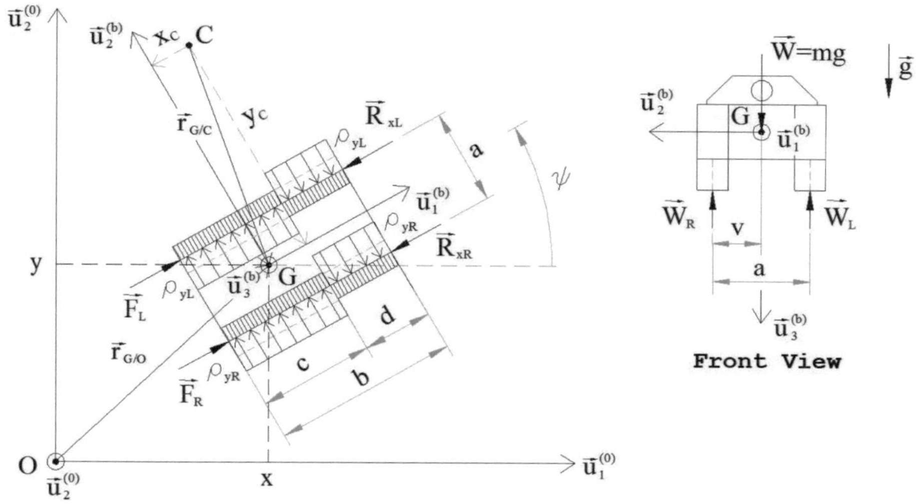

Figure 2.
Notable kinematic parameters of the tracked land vehicle and forces acting on the vehicle [8].

$\vec{W}_L$ and $\vec{W}_R$: weighting forces on the left and right tracks.

$\vec{R}_{XL}$ and $\vec{R}_{XR}$: longitudinal friction components acting on the left and right tracks.

$\vec{F}_L$and $\vec{F}_R$: actuation forces acting on the left and right tracks.

ρ_{xL} and ρ_{xR}: lateral friction force density acting on the left and right tracks.

For the tracked land vehicle, as u_x, u_y, and u_z denote the inputs and b_x, b_y, and b_ψ indicate the gravity and frictional force components, governing differential equations can be written in the following manner [8]:

$$\ddot{x} = u_x - b_x \tag{2}$$

$$\ddot{y} = u_y - b_y \tag{3}$$

$$\ddot{\psi} = u_\psi + b_\psi \tag{4}$$

where $u_x = u_F \cos(\psi)$, $u_y = u_F \sin(\psi)$, $u_F = (T_L + T_R)/(m r_S)$, $u_\psi = a(T_R - T_L)/(2 I_z r_S)$, $b_x = \sigma_{\dot{x}} \mu_x g \cos(\psi)$, $b_y = \sigma_{\dot{x}} \mu_x g \sin(\psi)$, and $b_\psi = \frac{\sigma_{\dot{\psi}} m g}{b I_z}\left[b \sigma_{\dot{x}}\left(v - \frac{a}{2}\right)\left(\frac{\mu_x}{\mu_y}\right) + x_C(c-d) - \frac{(c^2+d^2)}{2}\right]$.

Regarding these definitions, T_L and T_R stand for the actuation torques exerted by the power transmission gears on the left and right tracks; r_S and I_z represent the radius of its actuation gear and moment of inertia of the vehicle about the rotation axis indicated by the unit vector $\vec{u}_3^{(b)}$; and eventually μ_x and μ_y stand for the static friction coefficients between the tracks of the vehicle and surface on the lateral planes. Here, the symbols $\sigma_{\dot{x}}$ and $\sigma_{\dot{\psi}}$ are introduced as $\sigma_{\dot{x}} = \text{sgn}[\dot{x}\cos(\psi) + \dot{y}\sin(\psi)]$ and $\sigma_{\dot{\psi}} = \text{sgn}(\dot{\psi})$ where sgn$(\cdot)$ shows the signum function [8].

2.3 Description of the quadrotor

The schematic view of the quadrotor, the third system under consideration, and engagement geometry between the quadrotor and moving land platform are presented in **Figures 3** and **4**, respectively [8].

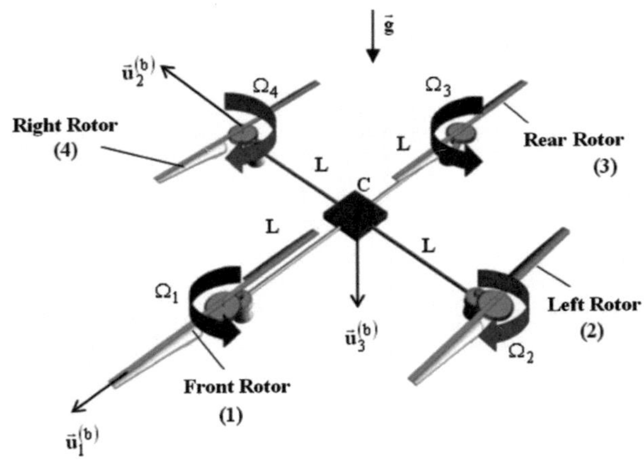

Figure 3.
Schematic view of the quadrotor [8].

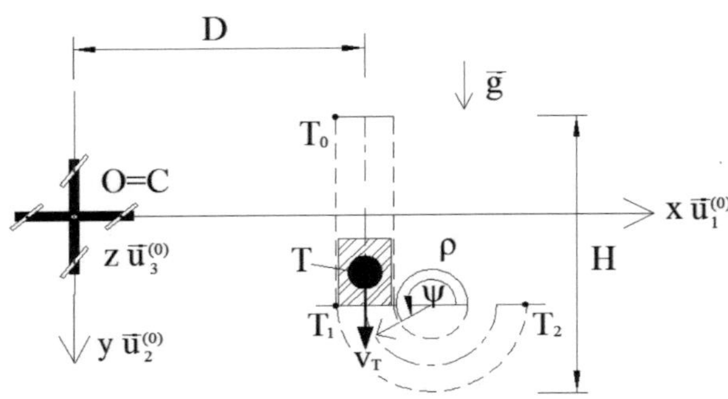

Figure 4.
Engagement geometry for the quadrotor [8].

As indicated in **Figure 3**, the front and rear rotors expressed as numbers 1 and 3, respectively, have rotations in positive sense around the axis represented by unit vector $\vec{u}_3^{(b)}$ of the body-fixed frame of the quadrotor, i.e., F_b, whose origin is attached at point C and whose axes are shown by unit vectors $\vec{u}_i^{(b)}$ (i = 1, 2, and 3), while the left and right rotors, i.e., numbers 2 and 4 rotors, rotate in negative sense [8].

In **Figure 3**, L and Ω_j (j = 1, 2, 3, and 4) stand for the distance between the center of rotation of each motor and point C and angular speed of the electromechanical actuator, i.e., electrical servomotor, used to move the propeller j, respectively. In addition to those parameters, the symbols in **Figure 3** can be listed as follows [8]:

T: target point on the moving platform for the quadrotor.
T_i: points at which the shape of the trajectory of the moving platform changes (i = 0, 1, and 2).
v_T: linear speed of point T on the moving platform.
ρ: radius of curvature of the trajectory of the moving platform.
ψ: rotation angle of the rounded tip portions of the trajectory of the moving platform.
H: total length of the trajectory of the moving platform.
D: perpendicular distance between the origin of F_O, i.e., point O, and the midline of the section of the trajectory of the moving platform closest to this point.

Considering the related kinematic and dynamic parameters of the quadrotor system, the dynamic model of the quadrotor can be set as follows using the relevant kinematic and dynamic parameters with angular position parameters of ϕ, θ, and ψ and translational position parameters of x, y, and z [8]:

$$\ddot{\phi} = -c_{\phi}\dot{\phi} + J_{\psi\theta}\dot{\psi}\dot{\theta} + J_{\theta}\Omega_e\dot{\theta} + u_{\phi} \tag{5}$$

$$\ddot{\theta} = -c_{\theta}\dot{\theta} + J_{\psi\phi}\dot{\psi}\dot{\phi} - J_{\phi}\Omega_e\dot{\phi} + u_{\theta} \tag{6}$$

$$\ddot{\psi} = -c_{\psi}\dot{\psi} + J_{\theta\phi}\dot{\theta}\dot{\phi} + u_{\psi} \tag{7}$$

$$\ddot{x} = -c_x\dot{x} + u_x \tag{8}$$

$$\ddot{y} = -c_y\dot{y} + u_y \tag{9}$$

$$\ddot{z} = -c_z\dot{z} - g + u_z \tag{10}$$

where $c_{\phi} = LK_{\phi}/J_x$, $J_{\psi\theta} = \left(J_y - J_z\right)/J_x$, $J_{\theta} = J_v/J_x$, $\Omega_e = \Omega_1 - \Omega_2 + \Omega_3 - \Omega_4$, $u_{\phi} = Lb\left(\Omega_2^2 - \Omega_4^2\right)/J_x$, $c_{\theta} = LK_{\theta}/J_y$, $J_{\psi\phi} = (J_z - J_x)/J_y$, $J_{\phi} = J_v/J_y$, $u_{\theta} = Lb\left(\Omega_1^2 - \Omega_3^2\right)/J_y$, $c_{\psi} = K_{\psi}/J_z$, $J_{\theta\phi} = \left(J_x - J_y\right)/J_z$, $u_{\psi} = d\left(-\Omega_1^2 + \Omega_2^2 - \Omega_3^2 + \Omega_4^2\right)/J_z$, $c_x = K_x/m$, $u_x = b\left(\Omega_1^2 + \Omega_2^2 + \Omega_3^2 + \Omega_4^2\right)[c(\psi)s(\theta)c(\phi) + s(\psi)s(\phi)]/m$, $c_y = K_y/m$, $u_y = b\left(\Omega_1^2 + \Omega_2^2 + \Omega_3^2 + \Omega_4^2\right)[s(\psi)s(\theta)c(\phi) - c(\psi)s(\phi)]/m$, $c_z = K_z/m$, and $u_z = b\left(\Omega_1^2 + \Omega_2^2 + \Omega_3^2 + \Omega_4^2\right)c(\theta)c(\phi)/m$.

In the definitions above, J_x, J_y, and J_z show the moment of inertia components of the quadrotor around the axes defined by unit vectors $\vec{u}_1^{(b)}$, $\vec{u}_2^{(b)}$, and $\vec{u}_3^{(b)}$, respectively. J_v represents the moment of inertia of each rotor about its axis of rotation, d indicates the drag factor, and K_{ϕ}, K_{θ}, K_{ψ}, K_x, K_y, and K_z stand for the aerodynamic moment and force components acting on the system in the roll, pitch, and yaw planes and along the longitudinal, lateral, and vertical planes, respectively. Furthermore, b indicates thrust factor of the motors [8].

3. Control systems

Since the guidance commands generated by the LHG law which is considered to make the motion planning of the autonomous systems so as to bring them to the specified point on the target trajectories are in terms of the linear velocity components of the mass centers of those systems, the main control variables of the systems are selected to be velocity components [8]. Also, for the sake of maintaining the stability of the systems, the gain matrices of the relevant control systems are continuously updated throughout the planned motion using the state information of target acquired by certain means like a camera.

3.1 Robotic arm control system

Because the guidance signals produced by the LHG law are in terms of the angle between the velocity vector of point P and lateral axis, the control variable of the robotic arm control system is chosen to be the joint speeds, i.e., $\dot{\theta}_1$ and $\dot{\theta}_2$ [6]. Since the main objective is the speed control of point P, the control system based on the joint speeds corresponds to an indirect control scheme. Regarding the difficulty in the measurement of the instantaneous speed values of point P along with the fact that joint angles and their rates can easily be acquired by means of the sensors put on the joints, the use of such an indirect control algorithm seems to be proper. In this sense, the linear position and velocity components of point P can be calculated using the measured joint positions and speeds via the loop closure equations, and consequently the trajectory of point P on the horizontal plane can be determined as a function of time [8].

Here, as $\dot{\theta}_{1d}$ and $\dot{\theta}_{2d}$ indicate the reference joint speeds and $\dot{\overline{\theta}}_d = \left[\dot{\theta}_{1d} \quad \dot{\theta}_{2d}\right]^T$ stands for the column matrix corresponds to these quantities, the error between the desired, i.e., reference, and actual joint speeds, i.e., $\overline{e}$, can be set as follows [8]:

$$\overline{e} = \dot{\overline{\theta}}_d - \dot{\overline{\theta}} \tag{11}$$

The relevant control rule can be written according to the computed torque method with the addition of the integral action to nullify the steady-state errors by regarding the error definition in Eq. (11) as follows [10]:

$$\overline{T} = \hat{M}\ddot{\overline{\theta}}_d + \hat{H}\dot{\overline{\theta}} + \hat{K}_p\overline{e} + \hat{K}_i\int \overline{e}\,dt \tag{12}$$

Here, as "T" represents the matrix transpose, $\overline{T} = [T_1 \quad T_2]^T$ demonstrates the torque column matrix for joint torques T_1 and T_2. Also, $\hat{M} = \hat{M}(\overline{\theta})$ and $\hat{H} = \hat{H}\left(\dot{\overline{\theta}}, \overline{\theta}\right)$ denote the inertia and centrifugal effect matrices, while $\hat{K}_p$ and $\hat{K}_i$ correspond to the proportional and integral gain matrices, respectively. As implied, the resulting control system becomes in PI (proportional plus integral) form [8].

The differential equation corresponding to the error dynamics of the control system for the robotic arm are obtained using Eqs. (11) and (12) in the following fashion:

$$\ddot{\overline{e}} + \hat{M}^{-1}\left(\dot{\hat{M}} + \hat{K}_p\right)\dot{\overline{e}} + \hat{M}^{-1}\hat{K}_i\overline{e} = \overline{0} \tag{13}$$

As ω_{ci} and ζ_{ci} denote the desired bandwidth and damping ratio of link i (i = 1 and 2), the error dynamics can be written for a second-order ideal system with two degrees of freedom as follows [10]:

$$\ddot{\overline{e}} + \hat{D}\dot{\overline{e}} + \hat{W}\overline{e} = \overline{0} \tag{14}$$

where $\hat{D} = \begin{bmatrix} 2\zeta_{c1}\omega_{c1} & 0 \\ 0 & 2\zeta_{c2}\omega_{c2} \end{bmatrix}$ and $\hat{W} = \begin{bmatrix} \omega_{c1}^2 & 0 \\ 0 & \omega_{c2}^2 \end{bmatrix}$.

Consequently, $\hat{K}_p$ and $\hat{K}_i$ are determined by equating Eq. (13) to Eq. (14) as follows [8]:

$$\hat{K}_p = \hat{M}\hat{D} - \dot{\hat{M}} \tag{15}$$

$$\hat{K}_i = \hat{M}\hat{W} \tag{16}$$

3.2 Tracked land vehicle control system

During the control of the tracked land vehicle for the angular variables, the angular position requirement for ψ should also be satisfied. For this purpose, a two-stage cascaded control scheme is constructed for the tracked land vehicle. In this algorithm, the outer loop is responsible of making linear velocity control in accordance with the guidance commands, while the inner loop makes the angular position control such that the orientation angle requirement arising during the linear velocity control is realized [8].

Here, the expressions which will be considered for the design of the linear velocity control system which is termed as "the primary control system" can be arranged in the forthcoming state space form from Eqs. (2) and (3) as $\overline{x}_p = [\dot{x} \quad \dot{y}]^T$, $\overline{u}_p$, and $\overline{b}_p$ stand for the column matrices for the state variables, system inputs, and gravity effect, respectively [8]:

$$\dot{\overline{x}}_p = -\overline{b}_p + \overline{u}_p \tag{17}$$

The control rule of the primary control system can be established using the computed torque control method according to the PI control action like the robotic arm control system so as to nullify the steady-state errors [8]:

$$\overline{u}_p = \dot{\overline{x}}_{pd} + \overline{b}_p + \hat{K}_{pp}\overline{e}_p + \hat{K}_{pi}\int \overline{e}_p\,dt \tag{18}$$

In the above equation, $\overline{x}_{pd}$, $\hat{K}_{pp}$, $\hat{K}_{pi}$, and $\overline{e}_p$ denote the column matrix for the desired inputs, gain matrix for the proportional control action, gain matrix for the integral control action, and column matrix for the error, respectively, for the primary control system. Here, $\overline{e}_p = \overline{x}_{pd} - \overline{x}_p$ is introduced.

Substituting Eq. (18) into Eq. (17) yields the next equation for the error dynamics of the primary control system [8]:

$$\ddot{\overline{e}}_p + \hat{K}_{pp}\dot{\overline{e}}_p + \hat{K}_{pi}\overline{e}_p = \overline{0} \tag{19}$$

As ω_{pi} and ζ_{pi} denote the desired bandwidth and damping ratio of link i (i = 1 and 2) for the primary control system, the error dynamics can be written for a second-order ideal system with two degrees of freedom as follows [8]:

$$\ddot{\overline{e}}_p + \hat{D}_p\dot{\overline{e}}_p + \hat{W}_p\overline{e}_p = \overline{0} \tag{20}$$

where $\hat{D}_p = \begin{bmatrix} 2\zeta_{p1}\omega_{p1} & 0 \\ 0 & 2\zeta_{p2}\omega_{p2} \end{bmatrix}$ and $\hat{W}_p = \begin{bmatrix} \omega_{p1}^2 & 0 \\ 0 & \omega_{p2}^2 \end{bmatrix}$.

Equating Eq. (19) to Eq. (20), the gain matrices are found for the primary control system as follows [8]:

$$\hat{K}_{pp} = \hat{D}_p \tag{21}$$

$$\hat{K}_{pi} = \hat{W}_p \tag{22}$$

The control rule for the angular control system called "the secondary control system" can be set using the computed torque method according to the PD (proportional plus derivative) control action in the following fashion [8]:

$$u_\psi = \ddot{\psi}_d - b_\psi + K_{sp}\,e_s + K_{sd}\,\dot{e}_s \tag{23}$$

In the above expression, b_ψ, ψ_d, K_{sp}, K_{sd}, and e_s represent the inertia gain for the vehicle, desired input variable for the secondary control system, gain of the proportional control action, gain of the derivative control action, and error, respectively. Also, $e_s = \psi_d - \psi$ is introduced.

Plugging Eq. (23) into Eq. (4), the error dynamics of the secondary control system is determined as follows [8]:

$$\ddot{e}_s + K_{sd}\,\dot{e}_s + K_{sp}\,e_s = 0 \tag{24}$$

As ω_s and ζ_s indicate the desired bandwidth and damping ratio for variable ψ for the secondary control system, the error dynamics can be written for a second-order ideal system with single degree of freedom as given below:

$$\ddot{e}_s + 2\zeta_s\,\omega_s\,\dot{e}_s + \omega_s^2\,e_s = 0 \tag{25}$$

Matching Eqs. (24) and (25) results in the forthcoming gains for the secondary control system [8]:

$$K_{sp} = \omega_s^2 \tag{26}$$

$$K_{sd} = 2\zeta_s\,\omega_s \tag{27}$$

3.3 Quadrotor control system

The same two-stage control system is designed for the quadrotor as that for the tracked land vehicle. For the primary control system, these expressions can be rearranged in the state space form by assigning the columns of the state variables and inputs of the system to be $\overline{x}_p = \begin{bmatrix}\dot{x} & \dot{y} & \dot{z}\end{bmatrix}^T$ and $\overline{u}_p = \begin{bmatrix}u_x & u_y & u_z\end{bmatrix}^T$, respectively, as letter "T" indicates the transpose operation as follows [8]:

$$\dot{\overline{x}}_p = -\hat{C}_p\,\overline{x}_p + \overline{u}_p \tag{28}$$

where $\hat{C}_p = \begin{bmatrix} c_x & 0 & 0 \\ 0 & c_y & 0 \\ 0 & 0 & c_z \end{bmatrix}$.

The control law of the primary control system including an integral action can be designed as per the computed torque method with PI action in the following manner [8]:

$$\overline{u}_p = \dot{\overline{x}}_{pd} + \hat{C}_p\,\overline{x}_p + \hat{K}_{pp}\,\overline{e}_p + \hat{K}_{pi}\int \overline{e}_p\,dt \tag{29}$$

In the expression above, $\overline{x}_{pd}$, $\hat{K}_{pp}$, $\hat{K}_{pi}$, and $\overline{e}_p$ stand for the desired input column, proportional gain matrix, integral gain matrix, and error column for the primary control system, respectively, and the definition $\overline{e}_p = \overline{x}_{pd} - \overline{x}_p$ is given.

Substituting Eq. (29) into Eq. (28), the error dynamics of the primary control system is obtained as follows [8]:

$$\ddot{\overline{e}}_p + \hat{K}_{pp}\,\dot{\overline{e}}_p + \hat{K}_{pi}\,\overline{e}_p = \overline{0} \tag{30}$$

As ω_{pi} and ζ_{pi} represent the desired bandwidth and damping ratio of the ith state variable (i = 1, 2, and 3) of the primary control system, respectively, the error dynamics of a second-order ideal system with three degrees of freedom can be described using the forthcoming expression [8]:

$$\ddot{\overline{e}}_p + \hat{D}_p\,\dot{\overline{e}}_p + \hat{W}_p\,\overline{e}_p = \overline{0} \tag{31}$$

where $\hat{D}_p = \begin{bmatrix} 2\zeta_{p1}\omega_{p1} & 0 & 0 \\ 0 & 2\zeta_{p2}\omega_{p2} & 0 \\ 0 & 0 & 2\zeta_{p3}\omega_{p3} \end{bmatrix}$ and $\hat{W}_p = \begin{bmatrix} \omega_{p1}^2 & 0 & 0 \\ 0 & \omega_{p2}^2 & 0 \\ 0 & 0 & \omega_{p3}^2 \end{bmatrix}$.

Matching Eqs. (30) and (31), the gain matrices appear for the primary control system as follows:

$$\hat{K}_{pp} = \hat{D}_p \tag{32}$$

$$\hat{K}_{pi} = \hat{W}_p \tag{33}$$

As the corresponding columns of the state variables and inputs of the system are shown to be $\overline{x}_s = [\phi \quad \theta \quad \psi]^T$ and $\overline{u}_s = [u_\phi \quad u_\theta \quad u_\psi]^T$, the matrix equality below is reached via Eqs. (5)–(7) for the orientation, or attitude, control system as the secondary control system [8]:

$$\ddot{\overline{x}}_s = \overline{b}_s + \overline{u}_s \tag{34}$$

where $\overline{b}_s = \left[-c_\phi\dot{\phi} + (J_{\psi\theta}\dot{\psi} + J_\theta\Omega_e)\dot{\theta} \quad -c_\theta\dot{\theta} + (J_{\psi\phi}\dot{\psi} - J_\phi\Omega_e)\dot{\phi} \quad -c_\psi\dot{\psi} + J_{\theta\phi}\dot{\theta}\dot{\phi}\right]^T$.

In a similar manner, the control law can be written according to the computed torque method with PD action for the secondary control system as follows [8]:

$$\overline{u}_s = \ddot{\overline{x}}_{sd} - \overline{b}_s + \hat{K}_{sp}\overline{e}_s + \hat{K}_{sd}\dot{\overline{e}}_s \tag{35}$$

Here, $\overline{x}_{sd}$, $\hat{K}_{sp}$, $\hat{K}_{sd}$, and $\overline{e}_s$ indicate the desired input column, proportional gain matrix, derivative gain matrix, and error column for the secondary control system, respectively. Also, the definition $\overline{e}_s = \overline{x}_{sd} - \overline{x}_s$ is made for the error term.

In the proposed entire control scheme, the desired values of ϕ and θ are calculated using u_x, u_y, and u_z inputs along with the parameter u_F by regarding the definitions within Eqs. (5)–(10). In other words, the reference inputs, or commands, to ϕ and θ are generated by the outer loop. On the other hand, the remaining angular position variable ψ is adjusted to be a constant value. That means its reference value is set as a fixed quantity. In this sense, the desired value of ψ is then specified to be zero as the decision on keeping the quadrotor with no angular motion in the yaw plane [11, 12].

Inserting Eq. (35) into Eq. (34), the error dynamics of the secondary control system is obtained as

$$\ddot{\overline{e}}_s + \hat{K}_{sd}\dot{\overline{e}}_s + \hat{K}_{sp}\overline{e}_s = \overline{0} \tag{36}$$

As ω_{si} and ζ_{si} denote the desired bandwidth and damping ratio of the ith state variable (i = 1, 2, and 3) of the secondary control system, the error dynamics of a second-order ideal system with three degrees of freedom can be described using the following equation [8]:

$$\ddot{\overline{e}}_s + \hat{D}_s\dot{\overline{e}}_s + \hat{W}_s\overline{e}_s = \overline{0} \tag{37}$$

where $\hat{D}_s = \begin{bmatrix} 2\zeta_{s1}\omega_{s1} & 0 & 0 \\ 0 & 2\zeta_{s2}\omega_{s2} & 0 \\ 0 & 0 & 2\zeta_{s3}\omega_{s3} \end{bmatrix}$ and $\hat{W}_s = \begin{bmatrix} \omega_{s1}^2 & 0 & 0 \\ 0 & \omega_{s2}^2 & 0 \\ 0 & 0 & \omega_{s3}^2 \end{bmatrix}$.

Equating Eqs. (47) and (48) to each other yields the gain matrices of the secondary control system as shown below [8]:

$$\hat{K}_{sp} = \hat{W}_s \tag{38}$$

$$\hat{K}_{sd} = \hat{D}_s \tag{39}$$

4. Engagement geometry

Since the motion of the moving land platform for the quadrotor, i.e., trajectory of the target point, is dealt with in the three-dimensional space, the lateral projection of the engagement geometry drawn for the quadrotor can be used for the engagements of the robotic arm and tracked land vehicle with their targets as well.

The engagement geometry between the quadrotor and the moving land platform is described in the horizontal and vertical planes separately. Thus, the vertical engagement between point C, the mass center of the quadrotor, and point T on the moving platform can be depicted as seen in **Figure 5**. The lateral engagement geometry between points C and T can be similarly constructed using the same expressions [8].

In **Figure 5**, v_C, γ_q, $r_{T/C}$, γ_t, and λ_p show the resultant speed of point C, orientation angle of v_C from the horizontal plane, relative position of point T with respect to point C, orientation angle of v_T from the horizontal axis, and orientation angle of $r_{T/C}$ from the horizontal axis. Here, the next equations are held for v_C, γ_t, and λ_p [8]:

$$v_C = \sqrt{\dot{x}_C^2 + \dot{y}_C^2 + \dot{z}_C^2} \tag{40}$$

$$\gamma_t = \begin{cases} -\pi/2, t_0 < t \leq t_1 \\ (\pi/2) - \psi, t_1 < t \leq t_2 \ \text{(rad)} \\ \pi/2, t > t_2 \end{cases} \tag{41}$$

$$\lambda_p = a\tan\left[(y_T - y_C)/(x_T - x_C)\right] \tag{42}$$

Similarly, the orientation angle of $r_{T/C}$ from the horizontal plane, i.e., λ_y, can be obtained as follows [8]:

$$\lambda_y = a\tan\left[(z_T - z_C)/(x_T - x_C)\right] \tag{43}$$

The final offset between points C and T at the end of the engagement, i.e., d_{miss}, is calculated using the next formula as t_F indicates the termination time [8]:

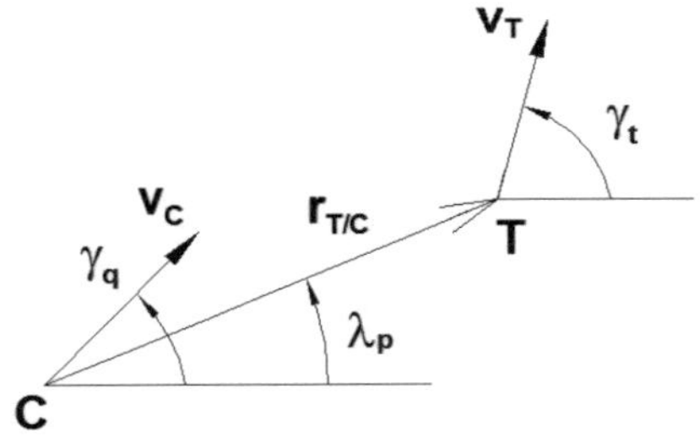

Figure 5.
Vertical engagement geometry between the mass center of the quadrotor and point defined on the moving platform [8].

$$d_{miss} = \sqrt{\Delta x^2(t_F) + \Delta y^2(t_F) + \Delta z^2(t_F)} \tag{44}$$

Here, as x_C, y_C, and z_C stand for the position components of point C and x_T, y_T, and z_T stand for the position components of point T on F_O, $\Delta x(t_F) = x_C(t_F) - x_T(t_F)$, $\Delta y(t_F) = y_C(t_F) - y_T(t_F)$, and $\Delta z(t_F) = z_C(t_F) - z_T(t_F)$.

5. Guidance law

In order for the considered point on the relevant system (point P for the robotic arm, point C for the tracked land vehicle, and point C for the quadrotor) to catch the desired point on the moving platform (point S for the robotic arm and point T for the tracked land vehicle and quadrotor), the guidance commands can be derived according to the LHG law for which the mechatronic system-target engagement geometry is depicted in **Figure 6** [13].

In **Figure 6**, M, T, and P stand for the mechatronic system, the target, and the predicted intercept point, respectively. Also, $\vec{v}_{Mactual}$ shows the velocity vector of the mechatronic system at the beginning of the guidance. The velocity vector of the mechatronic system in order to be on the collision triangle is then indicated by $\vec{v}_{Mideal}$. Once $\vec{v}_{Mactual}$ is turned into $\vec{v}_{Mideal}$, it means that the mechatronic system is on the collision triangle so as to collide the intended target at point P. In applying this method, unless the target has a velocity vector constant in both magnitude and direction, $\vec{v}_{Mideal}$ should be updated continuously in order to guarantee the intercept [13].

Regarding the LHG geometry expressed above verbally, the relevant guidance commands can be derived in terms of the orientation angles of v_C from the lateral and vertical axes (η_q^c and γ_q^c) as follows [6, 8]:

$$\eta_q^c = a \tan\left[(v_{Ty}\,\Delta t - \Delta y)/(v_{Tx}\,\Delta t - \Delta x)\right] \tag{45}$$

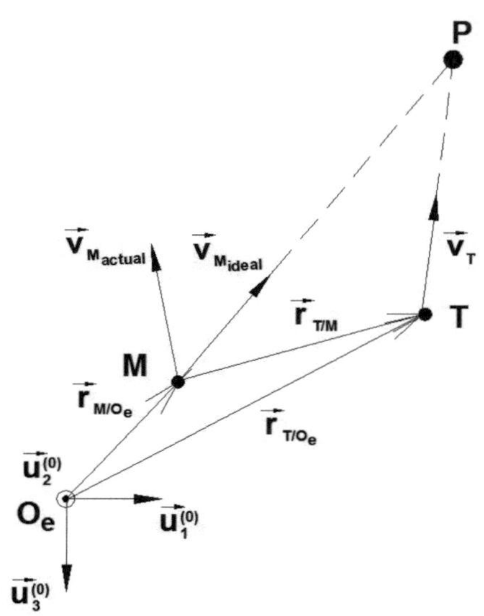

Figure 6.
Linear homing guidance law geometry [13].

$$\gamma_q^c = a\tan\left[\frac{\Delta z - v_{Tz}\,\Delta t}{(v_{Tx}\,\Delta t - \Delta x)\,\cos\left(\eta_q^c\right) + (v_{Ty}\,\Delta t - \Delta y)\,\sin\left(\eta_q^c\right)}\right] \tag{46}$$

where $\Delta x = x_C - x_T$, $\Delta y = y_C - y_T$, and $\Delta z = z_C - z_T$.

In Eqs. (45) and (46), the components of the linear velocity vector of point T whose amplitude is v_T, i.e., v_{Tx}, v_{Ty}, and v_{Tz}, are defined in the following equations [6, 8]:

$$v_{Tx} = v_T\,\cos\left(\gamma_t\right) \tag{47}$$

$$v_{Ty} = v_T\,\sin\left(\gamma_t\right) \tag{48}$$

$$v_{Tz} = 0 \tag{49}$$

Furthermore, the remaining time till the end of the engagement, i.e., Δt, is formulated below:

$$\Delta t = \left[\sqrt{\sigma^2 + \left(v_C^2 - v_T^2\right)\Delta r^2} - \sigma\right] / \left(v_C^2 - v_T^2\right) \tag{50}$$

In the above equation, $\sigma = v_{Tx}\,\Delta x + v_{Ty}\,\Delta y + v_{Tz}\,\Delta z$ and $\Delta r^2 = \Delta x^2 + \Delta y^2 + \Delta z^2$.

In this work, it is assumed that the speed and orientation parameters of the moving target are obtained by processing the data acquired by the camera on the system under control. Since the control inputs of the designed control system are linear velocity components, the guidance commands given in Eqs. (45) and (46) should be expressed in terms of the linear velocity parameters for realization. This transformation can be done by writing the velocity vector of point C with amplitude v_C in terms of its components on F_0 as follows [6, 8]:

$$\overline{x}_{pd} = v_C \begin{bmatrix} \cos\left(\eta_q^c\right)\,\cos\left(\gamma_q^c\right) \\ \sin\left(\eta_q^c\right)\,\cos\left(\gamma_q^c\right) \\ -\sin\left(\gamma_q^c\right) \end{bmatrix} \tag{51}$$

Parameter	Numerical value
a_1 and a_2	1.25 m
d_1 and d_2	0.625 m
m_1 and m_2	10 kg
I_{c1} and I_{c2}	1.302 kg·m^2
b_1 and b_2	0.001 N·m·s/rad
ω_{c1} and ω_{c2}	10 Hz
ζ_{c1} and ζ_{c2}	0.707
L	2 m
ρ	0.5 m
d	1.5 m

Table 1.
Numerical values used in the simulations for the robotic arm [8].

6. Computer simulations

The system trajectories acquired from the computer simulations performed in accordance with the numerical values given in **Tables 1–3** for the robotic arm, tracked land vehicle, and quadrotor are submitted in **Figures 7–10** along with the corresponding target motions. Having constructed the engagement geometry between the mechatronic system under consideration and target, the LHG law is applied for these situations. In the simulations, disturbance effects due to the nonlinear friction characteristic and noise on the sensors on the joints are assumed

Parameter	Numerical value
a	2.5 m
b	4 m
v	1.25 m
m	25,000 kg
I_z	60,000 kg·m^2
r_s	0.3 m
μ_x and μ_y	0.4
ω_{p1} and ω_{p2}	10 Hz
f_s	30 Hz
ζ_{p1}, ζ_{p2}, and ζ_s	0.707

Table 2.
Numerical values used in the simulations for the tracked land vehicle [8].

Parameter	Numerical value
L	0.25 m
b	5×10^{-5} N·s^2
d	1×10^{-6} N·m·s^2
m	2 kg
I_x and I_y	0.2 kg·m^2
I_z	0.3 kg·m^2
J_v	1×10^{-3} kg·m^2
K_x, K_y, and K_z	0.01 N·s/m
K_ϕ, K_θ, and K_ψ	0.012 N·s/m
ω_{pi}	5 Hz
ω_{si}	15 Hz
ζ_{pi} and ζ_{si}	0.707
H	50 m
ρ	15 m
D	50 m
Solver step	1 ms

Table 3.
Numerical values used in the simulations for the quadrotor [8].

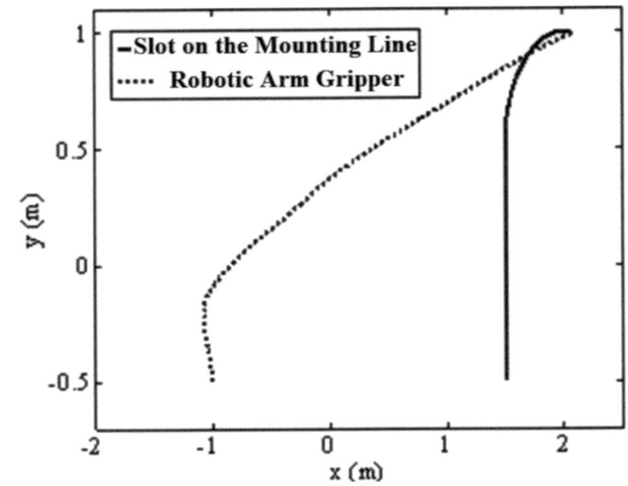

Figure 7.
Engagement geometry of the robotic arm with the mounting line.

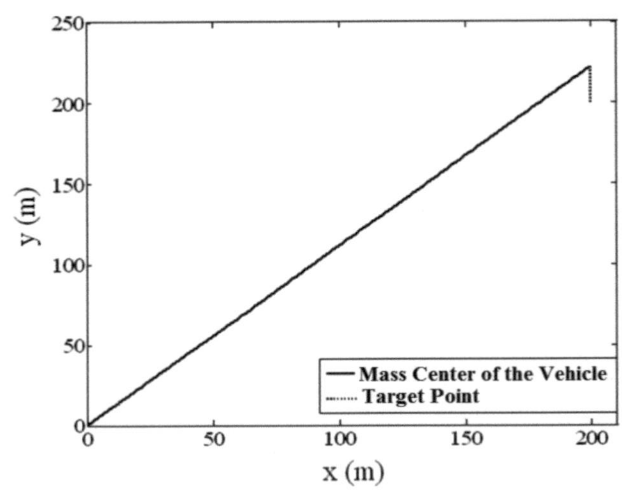

Figure 8.
Engagement geometry of the tracked land vehicle with the constant speed target point.

to randomly change within the intervals of ± 10 N·m and $\pm 1 \times 10^{-3}$ rad for the robotic arm. Also, it is regarded that the angular and linear dynamics of the quadrotor are subjected to random disturbing moment and force with maximum amplitudes of 50 N·m and 100 N, respectively. The simulations of the tracked land vehicle are made on nominal operating conditions [8].

7. Conclusion

As a result of the performed computer simulations, it is shown that the considered autonomous mechatronic systems, i.e., the robotic arm, tracked land vehicle, and quadrotor, can catch the specified target points by regarding the LHG law. Although only one engagement case is presented for each of the systems above, the same result is attained for different situations, too. In this scene, one of the most

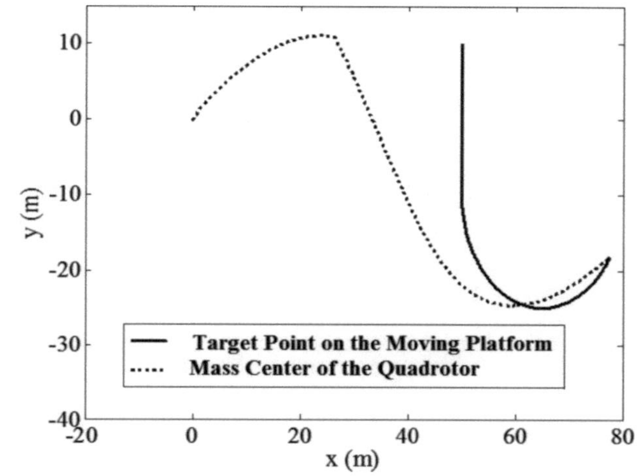

Figure 9.
Horizontal engagement geometry of the quadrotor with the platform.

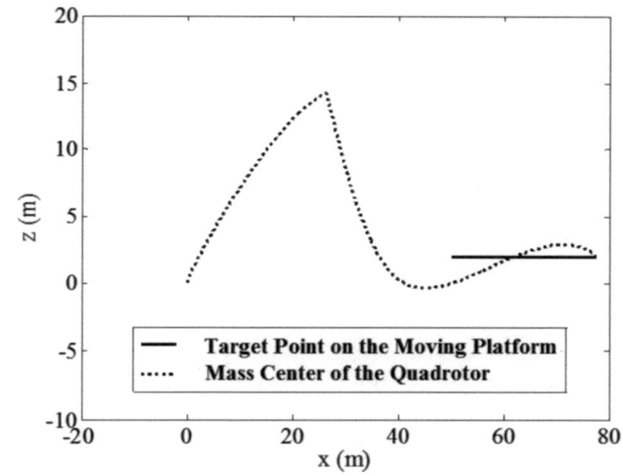

Figure 10.
Vertical engagement geometry of the quadrotor with the platform.

important considerations is the capacity of the actuators of the autonomous systems. Namely, if the maximum force or torque, hence maximum current, level of the actuators (electric motors) does not satisfy the requirements arising due to the planned motion profile, then the relevant system cannot track the target as planned. In general, it can be concluded that the motion planning of mechatronic systems including the service robots can be made against predefined target points by choosing a convenient guidance law.

Author details

Bülent Özkan
The Scientific and Technological Research Council of Turkey, Defense Industries Research and Development Institute (TÜBİTAK SAGE), Turkey

*Address all correspondence to: bozkan37@gmail.com

References

[1] Debnath SSK, Omar R, Latip NBA. A review on energy efficient path planning algorithms for unmanned air vehicles. Computational Science and Technology, Lecture Notes in Electrical Engineering. Singapore: Springer Nature Singapore Pte Ltd.; 2019;**481**:523-532

[2] Mahmoud Zadeh S, Powers DMW, Bairam Zadeh R. State-of-the-art in UVs' autonomous motion planning. Autonomy and Unmanned Vehicles, Cognitive Science and Technology. Singapore: Springer Nature Singapore Pte Ltd.; 2019. pp. 31-40

[3] Du X, Li X, Li D, Dai B. Path planning for autonomous vehicles in complicated environments. In: 2016 IEEE International Conference on Vehicular Electronics and Safety (ICVES); 2016. pp. 54-60

[4] Minh VT, Pumwa J. Feasible path planning for autonomous vehicles. Mathematical Problems in Engineering. 2014;**2014**:1-12

[5] Zarchan P. Tactical and Strategic Missile Guidance. 2nd ed. USA: Progress in Astronautics and Aeronautics; 1994

[6] Özkan B, Özgören MK, Mahmutyazıcıoğlu G. Comparison of the acceleration- and angle-based guidance laws for a short-range air-to-air missile (in Turkish). In: TOK2008-Automatic Control National Meeting. İstanbul, Turkey: İstanbul Technical University; 2008

[7] Kunwar F, Benhabib B. Advanced predictive guidance navigation for mobile robots: A novel strategy for rendezvous in dynamic settings. International Journal on Smart Sensing and Intelligent Systems. 2008;**1**(4):858-890

[8] Özkan B. Investigation of the usefulness of the linear homing guidance law for the motion planning of mechatronic systems. In: 17th International Carpathian Control Conference (ICCC 2016), Tatranská Lomnica, Slovakia; 2016

[9] Tsay TS. Guidance and control laws for quadrotor UAV. WSEAS Transactions on Systems and Control. 2014;**9**(1):606-613

[10] Ogata K. Modern Control Engineering. 2nd ed. New Jersey, USA: Prentice-Hall International Editions; 1990

[11] Voos H, Nourghassemi B. Nonlinear control of stabilized flight and landing for quadrotor UAVs. In: Proceedings of the 24th Annual Meeting of the European Institute for Applied Research (IAR'2009). Góra, Poland: Zielona; 2009

[12] Zhang D, Qi H, Wu X, Xie Y, Xu J. The quadrotor dynamic modeling and indoor target tracking control method. Mathematical Problems in Engineering. 2014;**2014**:1-9

[13] Özkan B. Dynamic modeling, guidance, and control of homing missiles [PhD dissertation]. Ankara, Turkey: Middle East Technical University; 2005

Chapter 5

Conversion of a Conventional Wheelchair into an Autonomous Personal Transportation Testbed

Volkan Sezer, Rahman Salim Zengin, Hosein Houshyari and Murat Cenk Yilmaz

Abstract

Personal transportation is the act of transporting an individual by using a small, low-speed vehicle. It is a very hot research topic both in industry and academia. There are many different types of personal transportation vehicles, and wheelchairs are one of them. Autonomous driving is another very popular subject that is applicable to the personal transportation vehicles. Autonomous personal transportation vehicles are good examples of service robotics applications. In this study, conversion procedure of a conventional electric wheelchair into an autonomous personal transportation testbed and the application of some basic autonomous driving algorithms on the developed testbed are explained. In literature, there are several studies providing information on wheelchairs' autonomy but not deep information about the conversion itself. In this paper, the conversion process is investigated in detail, under two main sections. The first part is by-wire conversion, which allows the wheelchair to be controlled via computer commands. The second part includes the studies on sensors, computational system, and human interface. After making such modifications on wheelchair, fundamental algorithms required for autonomy, such as mapping and localization, are implemented successfully. The results are promising for the usage of the developed system as a testbed for examining new autonomous algorithms and evaluating the performance of the perceptional/computational components.

Keywords: autonomous wheelchair, navigation, localization, drive by wire, personal transportation

1. Introduction

Traveling autonomously from one place to another is an extremely important subject, which has been a dream for years and studied a lot in recent years. The research on the subject generally focuses on the autonomy of passenger cars going into traffic. The first serious results of the studies on this subject came from the race named Darpa Grand Challenge, which was organized by the organization "Defense Advanced Research Projects Agency (DARPA)," which carries out the United States' advanced technology defense projects, in 2004, and then continued in 2005 and 2007, respectively [1]. After DARPA competitions, lots of autonomous studies have been done all over the world such as in [2–4].

IntechOpen

Autonomous traveling is possible and a serious need, not only in the way that passenger cars perform in traffic, but also in closed environments where the use of Global Positioning System (GPS) is not possible and where the density of people is high. People may prefer to go somewhere autonomously in a closed environment even if there is no walking disability. Wheelchairs are very suitable vehicles for this kind of personal transportation. Also, for some people with disabilities, wheelchairs are the only option to get from one place to another. Unfortunately, many people with disabilities lack the ability to safely use their wheelchairs. This situation poses serious risks for them, primarily for the people around them and other environmental elements. In line with all these needs, in this paper, design and development of a fully autonomous smart wheelchair is explained.

In literature, single-person autonomous/semi-autonomous vehicle studies are generally carried out on the wheelchair platform. The wheelchair named TetraNauta in [5] was developed at the University of Seville between 1998 and 2004, and it operates in a known map autonomously. The study shown in [6] explains the smart wheelchair project that started in 2004 at the Massachusetts Institute of Technology. In this study, an efficient, socially acceptable autonomous tour-following behavior was developed. In another work about autonomous wheelchairs [7], RGB-D camera was used as the main perception sensor, and the map of the environment is constructed from this. Ref. [8] provides information on the development of a robot operating point (ROS)-based autonomous wheelchair that will work indoors. Ref. [9] mentions an autonomous wheelchair that uses only light detection and ranging (LIDAR) as its environmental measurement unit. This study is carried out using the ROS platform. In [10, 11], a semi-autonomous wheelchair was designed where the chair was controlled via head movements. In [12], the aim is to estimate the chair pose from video data by using machine learning methods using artificial neural networks. Another autonomous wheelchair study [12] provides the results about navigation in cluttered environments without explicit object detection and tracking.

All these studies use autonomous/semi-autonomous wheelchairs, which are converted from a conventional wheelchair platform. They provide information about their wheelchairs' autonomy but not deep information about the conversion itself. The conversion process of autonomous systems is explained in some papers for autonomous automobiles [13, 14] but not very detailed for wheelchairs. The dimensions, velocity capabilities, and differential drive architecture make wheelchairs different than standard automobiles. In this paper, we explain how to convert a conventional wheelchair into an autonomous one which is aimed to be used as a testbed for advanced autonomous algorithms. Besides, the results of localizing and mapping algorithms applied on this testbed are illustrated at the end of the work.

2. By-wire conversion

The conventional wheelchair used in this work is a standard differential drive platform with 2 × 500 W electric motors and a 24 V-40 Ah lead acid battery. Its weight (without person) is 80 kg with the dimensions of 120 cm × 65cm × 100cm. After some effort, now it is able to be driven by the outputs of the autonomy algorithms. **Figure 1** shows the conventional version of the wheelchair which is controlled by a joystick.

The original electric motor driver on this wheelchair is communicating with the joystick via a non-standard protocol, and it was a black box for us since there is no information about how it operates. For this reason, as a first step toward a

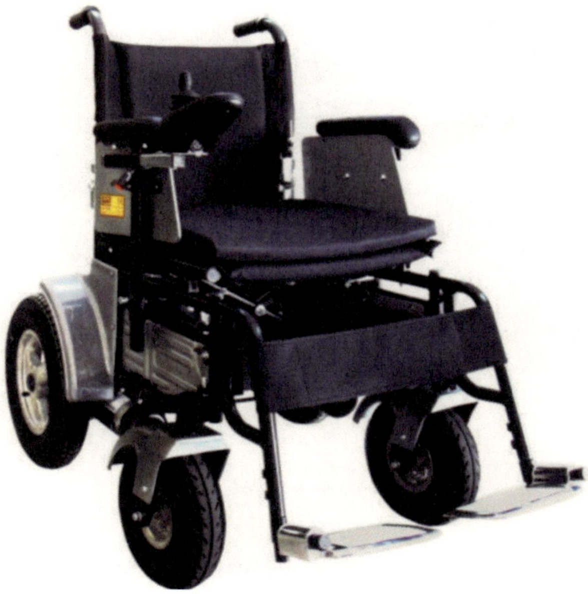

Figure 1.
Conventional electric wheelchair used as a base system for conversion.

full autonomy, the motor drivers should be replaced by another one that is able to communicate via a standard protocol and well documented. Therefore, we replaced the motor driver of each motor with a standard brushed DC motor driver "Roboteq MDC2230," which is a dual channel driver and is able to provide 50 A current continuously for each channel.

In robotic applications, electric motors are generally used in "velocity control mode" instead of "torque mode." For this reason, the motor driver needs to get the actual velocity value as a reference signal to be tracked. Hence, it is mandatory to measure the motor velocity for a feedback controller. For this reason, additional encoders (Atek-ARCB50360HPL33MY8FZ) for each electric motor are included to the system. The joystick is also removed since it is not necessary anymore in the autonomous concept. As a result, we mounted a motor driver that includes two separated drivers in the same box and two encoders. The motor driver box and one of the encoders are shown in **Figure 2**.

According to the motor driver's data interface, the driver is able to communicate via serial port, which is suitable for our computational system. Since it is planned to use the robot operating system (ROS) [15] to develop and implement the autonomous algorithms, a ROS node is also written to convert ROS commands to serial port data packages using "Rosserial." Rosserial is a protocol for wrapping standard ROS serialized messages and multiplexing multiple topics and services over serial port. As a result, by mounting additional components (two motor drivers and two encoders) and writing a ROS node, the wheelchair is converted to a drive-by-wire system, which means it can take reference velocity commands from ROS and applies it successfully.

Finally, we added an on-off button in order to open and shut down the whole system. In addition, an emergency button is mounted near the seat to be reached by the human easily. The emergency button is connected to the appropriate input of the motor driver which cuts down the energy to motors immediately when emergency signal comes. The on-off and emergency buttons are shown in **Figure 2**.

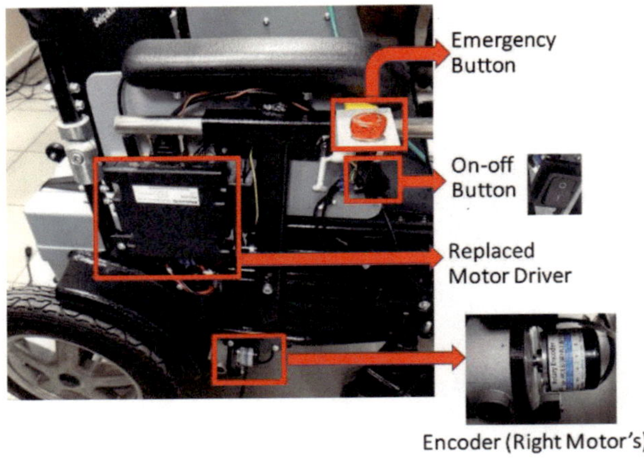

Figure 2.
Motor driver, encoder, on-off, and emergency buttons mounted.

3. Sensors, computational system, and human-machine interface

It is planned to use the system as a research and development testbed at the end of the conversion. For this reason, several sensors and computational hardware with different characteristics are mounted to the chair. These devices will be used for performance comparison in the future. **Tables 1** and **2** show the list of the sensors for perception and the computational hardware added to the autonomous wheelchair testbed. More information about these components is illustrated in the next subsections.

3.1 Perception system

In this study, we have three main perception sensors: one RGB-Depth camera (Intel Real Sense—D435i) that provides point cloud data with 87° × 58° × 95° field of view and 10 m range; one industrial relatively expensive 2D Lidar (SICK LMS151) that has 270° field of view, maximum 50 m range, and 50 Hz scan rate; and one low-cost Lidar (RPLIDAR-A2M6) that has 360° field of view, maximum 18 m range, and 15 Hz scan rate.

The aim is to analyze the performance of these different characteristic sensors on global mapping, local mapping, tracking, and localization. Some of these sensors can be removed from the system or can be used together depending on their perception performance. The angle of RGB-D camera can be adjusted manually. **Figure 3** shows these sensors that are mounted on the chair, using appropriate mechanical parts which are designed for the chair specifically.

3.2 Computational system

In our design, three computational subsystems are added to the system where two of them are planned for autonomous driving algorithms based on ROS platform. The first option is an embedded computer JETSON-TX2 from Nvidia. This low power consumption board is built around NVIDIA Pascal™-family graphics processing unit (GPU) and very suitable for matrix operations used in deep learning. Another option is a low power consumption small size embedded computer, AAEON Upboard. It has Intel® Atom™ x5 Z8350 Processor (Cherry Trail) of 64 bits up to 1.92 GHz. The final computational board is going to be used as a real-time

Device	Summary
INTEL REAL-SENSE—D435i	RGB-Depth camera with 87° × 58° × 95° field of view and 10 m range
SICK LMS151	2D Lidar with field of view of 270° and maximum 50 m range. Scanning frequency is 25–50 Hz
RPLIDAR-A2M6	2D Lidar with field of view of 360° and maximum 18 m range. Scanning frequency is 5–15 Hz

Table 1.
Perception sensors used in the testbed.

Device	Summary
JETSON-TX2	GPU 256 NVIDIA CUDA cores CPU Dual-Core NVIDIA Denver 2 64-Bit CPU Quad-Core ARM® Cortex®-A57 MPCore
AAEON Upboard	Intel® Atom™ ×5 Z8350 Processor (Cherry Trail) of 64 bits up to 1.92 GHz
ST B-F446E-96B01A	Based on the STM32F446 microcontroller, including 9-axis accelerometer/gyroscope/ magnetometer

Table 2.
Computational hardware used in the testbed.

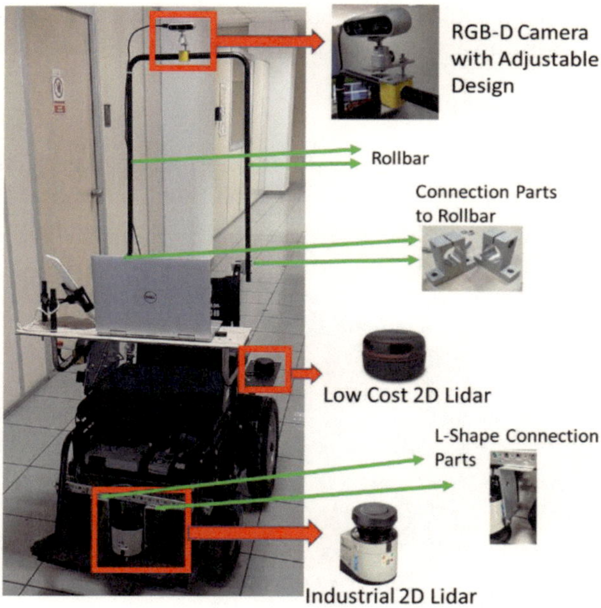

Figure 3.
Perception sensors on wheelchair.

low-level controller if the performance of the motor drivers' velocity control is not enough. This embedded board is B-F446E-96B01A from ST and is based on the STM32F446 microcontroller, including 9-axis accelerometer/gyroscope/ magnetometer. Similar to the aim of using different perception sensors, finding the optimum combination of computational hardware for autonomous algorithms is another objective for the project. Since it is not possible to know the computational power requirements before the decision of autonomous driving algorithms, we will analyze the performance of each hardware. At the end, one of these boards can be removed from system or can be used together depending on their performance. **Figure 4** illustrates the computational hardware mounted in a drawer, which is designed for the chair specifically. The gray protection boxes for Jetson TX2 and Upboard, which are produced by 3D printer, can be seen from **Figure 4**.

3.3 Power distribution

The main energy source of the system consists of two serial connected lead acid batteries. The main voltage level for the traction motors is 24 V. On the other hand, there are several other components for autonomous operation. In order to provide the appropriate voltage level for each sensor and computational hardware, DC/DC converters are used. **Figure 5** illustrates the power distribution scheme, including voltage levels, of the wheelchair components. An additional electric circuit is designed for power distribution using appropriate connectors fuses. DC/DC converters and the designed circuit can be seen from **Figure 4**.

3.4 Communication structure

Until now, all the electronic components have been shown and their purpose of use has been explained. In order to use all these devices properly, a secure communication network needs to be constructed. Each component has different interface in the system. For this reason, there are five different communication protocols such as Ethernet, USB 2.0, RS232, USB 3.1, and analog signals. **Figure 6** illustrates the overall communication architecture of the developed autonomous wheelchair.

3.5 Human-machine interface

The final requirement for the autonomous wheelchair is a user interface that allows the human user to send the desired goal point on map. After this selection, the wheelchair plans its trajectory and tracks the path continuously. Instead of

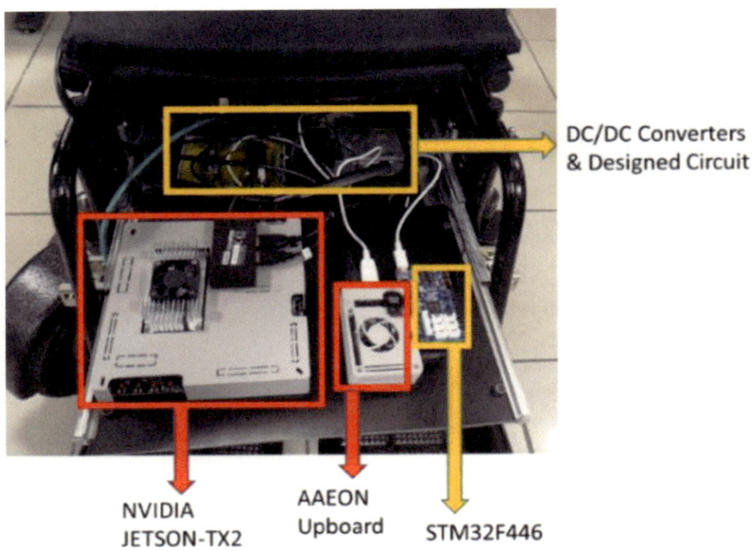

Figure 4.
Computational hardware of autonomous wheelchair (drawer is open).

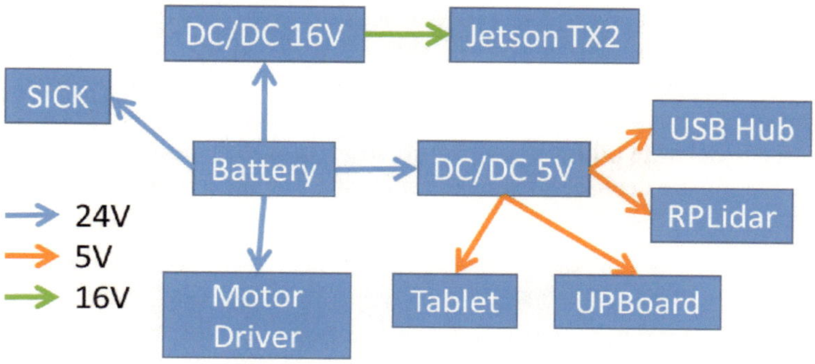

Figure 5.
Power distribution of autonomous wheelchair.

Figure 6.
Overall communication architecture of the system.

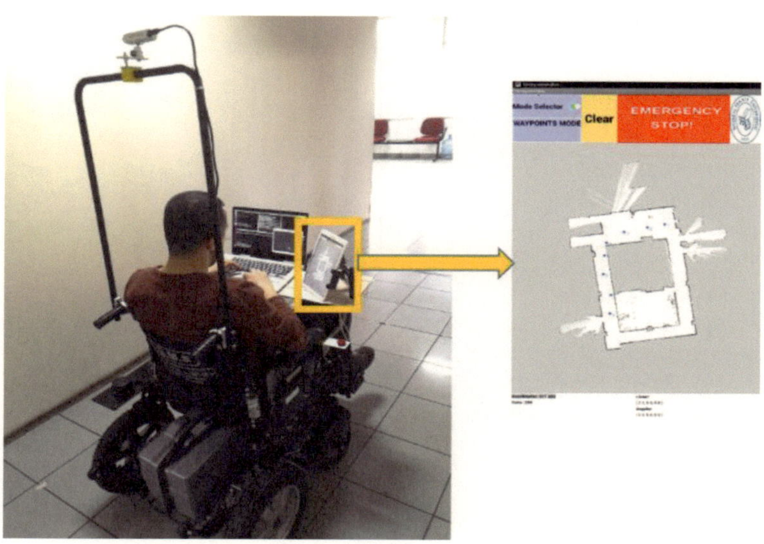

Figure 7.
Rear view of autonomous wheelchair and screenshot of the human-machine.

providing only the goal point, the user can set the waypoints to be tracked, by touching the desired coordinates on the map. Additional features of this interface are to send an emergency signal to the chair and to see the critical information such as wheelchair's velocity and actual position on map. The interface software is designed on a touchable tablet PC. Assembled form of the tablet on wheelchair and a screenshot from the designed interface software are shown in **Figure 7**.

4. Experiments

After the modifications mentioned previously, now, we have a testbed that is being used for testing the autonomous driving algorithms such as simultaneous localization and mapping (SLAM), trajectory planning, trajectory tracking, and low-level controllers. Until now, using the designed testbed, the map of the test environment is constructed using "gmapping" library [16] in ROS development environment. **Figure 8** shows a part of the environment where the mapping and localization algorithms are tested.

Figure 8.
Real test environment.

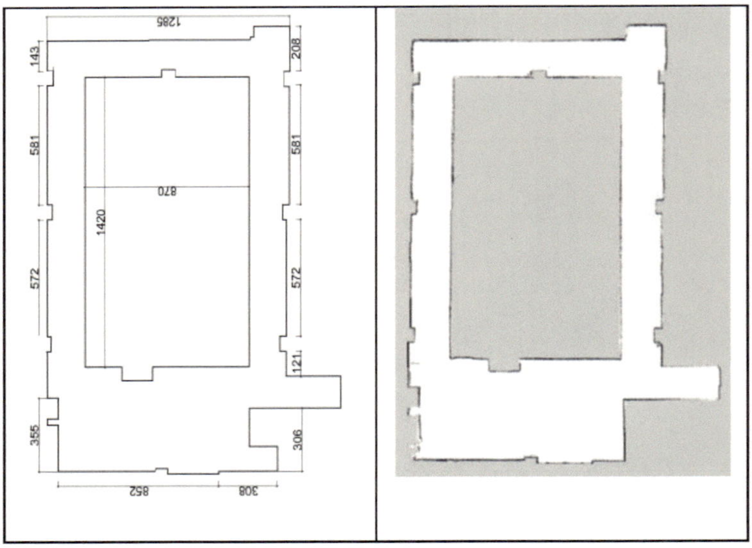

Figure 9.
Real map drawn using AutoCAD® software (left) and constructed map by SLAM (right).

Figure 9 illustrates the real map of the environment which is drawn using AutoCAD® software using accurate manual measurements and the constructed map drawn by the autonomous wheelchair using SLAM. The famous SLAM package of ROS named as "gmapping" is used for map construction. The map is based on SICK LMS-151 Lidar and the odometry information. As it is seen from **Figures 10** and **11**, the constructed map and the real map are almost same. Since the office doors are closed in the AutoCAD version, the constructed map for comparison is prepared in the same condition. According to the structural similarity index method (SSIM) [17], when the window size is taken as 5 × 5 pixels, the similarity of two maps is 94%.

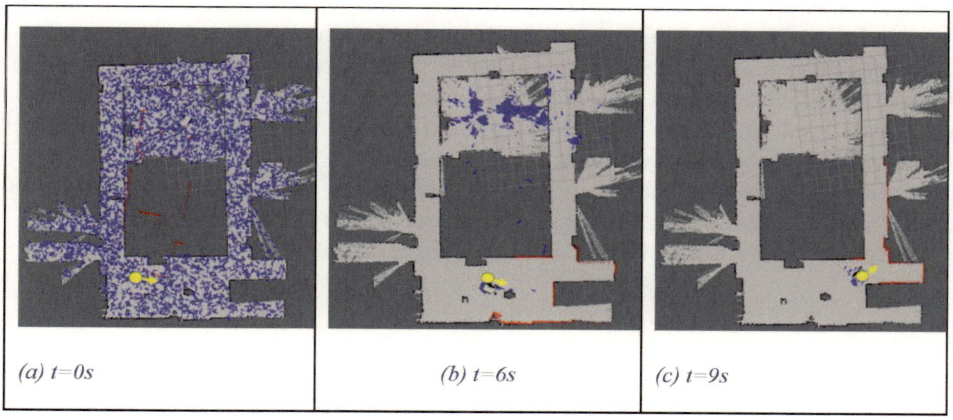

Figure 10.
Adaptive particle filter localization application (blue arrows: particle poses, yellow: real pose).

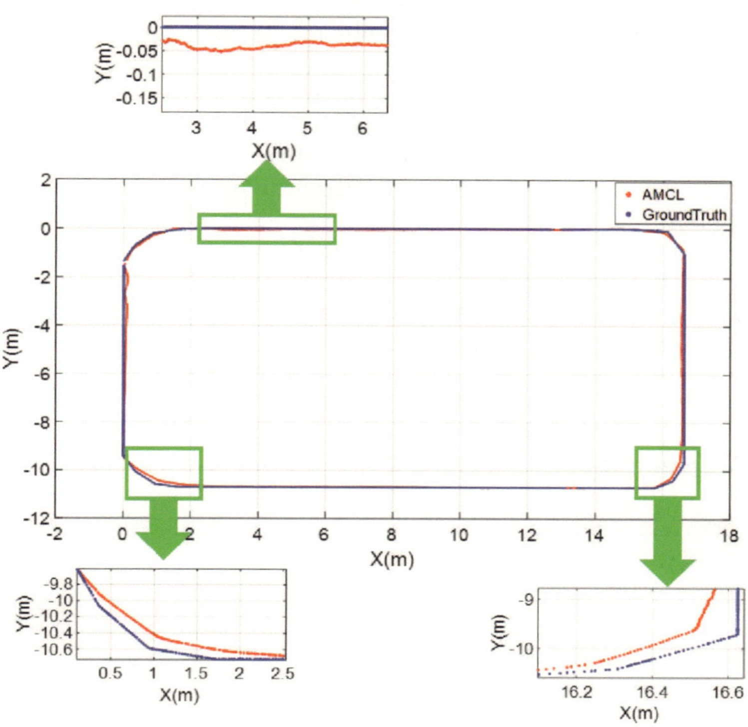

Figure 11.
Localization performance of the wheelchair.

Another application is global localization of the wheelchair. In order to plan a path and track it, an autonomous agent must be aware of its pose in the map. Adaptive Monte Carlo Localization (AMCL), which is based on particle filter, is used for pose estimation. The localization algorithm is run separately after the map construction. This time we use a map that was constructed when the doors are open since the localization tests were done in open-door condition. **Figure 10** shows the poses of particles (blue) and the real pose (yellow) during wheelchair's motion. It should be noted that the wheelchair is driven manually for both mapping and localization.

In **Figure 10**, the figure in the left shows the beginning phase, and all the particles are scattered into environment uniformly. No prior information is provided as initial pose. The figure in the middle illustrates that the particles concentrated on places coherent with the Lidar and odometry measurements. Finally in the right figure, it is shown that particles are gathered around the real position of the wheelchair after 9 seconds. The real position of the wheelchair is calculated by the manual measurements from the wheelchair.

Figure 11 shows the localization performance of the wheelchair in the same real environment. It is observed that the average localization error after the points are converged is below 10 cm, which is acceptable for our future studies.

5. Conclusion

In this study, the conversion procedure of a conventional electric wheelchair into an autonomous personal transportation testbed is described in detail. The conversion process is investigated under two main sections. The first part is by-wire conversion that allows the wheelchair to be controlled via digital commands. The second part includes the studies on the sensors, computational system and human-machine interface. Sensors and computational hardware that have different characteristics are included in the system for comparison and optimization of their performance in the future. The platform was tested using SLAM and AMCL algorithms for mapping and localization successfully. It is observed that the constructed map using industrial LIDAR and odometry data is almost the same with the real map and localization performance is acceptable for the next studies.

In the future, we plan to use the wheelchair as a research platform to further improve the autonomous personal transportation algorithms that will be used in narrow and cluttered environments. Comparison of localization and mapping performance of different methods and sensors in such environments will be studied as well.

Acknowledgements

This work was supported by the Turkish Scientific and Technological Research Council (TUBITAK) under project no. 118E809.

Author details

Volkan Sezer*, Rahman Salim Zengin, Hosein Houshyari and Murat Cenk Yilmaz
Istanbul Technical University, Istanbul, Turkey

*Address all correspondence to: sezerv@itu.edu.tr

References

[1] Buehler M, Iagnemma K, Singh S, editors. The 2005 DARPA Grand Challenge: The Great Robot Race. Vol. 36. Springer; 2007

[2] Broggi A, Medici P, Zani P, Coati A, Panciroli M. Autonomous vehicles control in the VisLab intercontinental autonomous challenge. Annual Reviews in Control. 2012;**36**(1):161-171

[3] Poczter SL, Jankovic LM. The Google car: Driving toward a better future? Journal of Business Case Studies (JBCS). 2013;**10**(1):7

[4] Sezer V, Bandyopadhyay T, Rus D, Frazzoli E, Hsu D. Towards autonomous navigation of unsignalized intersections under uncertainty of human driver intent. In: 2015 IEEE/RSJ International Conference on Intelligent Robots and Systems (IROS). Hamburg, Germany: IEEE; 2015

[5] Diaz SV, Rodriguez CA, Rio FDD, Balcells AC, Muniz DC. TetraNauta: A intelligent wheelchair for users with very severe mobility restrictions. Proceedings of the International Conference on Control Applications. Vol. 2. pp. 778-783

[6] Hemachandra S, Kollar T, Roy N, Teller S. Following and interpreting narrated guided tours. In: 2011 IEEE International Conference on Robotics and Automation. Shanghai, China: IEEE; 2011

[7] Baklouti E, Amor NB, Jallouli M. Autonomous wheelchair navigation with real time obstacle detection using 3D sensor. Automatika. 2016;**57**(3):761-773

[8] Li Z, Xiong Y, Zhou L. ROS-based indoor autonomous exploration and navigation wheelchair. In: 2017 10th International Symposium on Computational Intelligence and Design (ISCID). Hawaii, USA: IEEE; 2017

[9] Grewal H, Matthews A, Tea R, George K. LIDAR-based autonomous wheelchair. In: 2017 IEEE Sensors Applications Symposium (SAS). New Jersey, USA: IEEE; 2017

[10] Demir M, Sezer V. Design and implementation of a new speed planner for semiautonomous systems. Turkish Journal of Electrical Engineering & Computer Sciences. 2018;**26**(2):693-706

[11] Mavus U, Sezer V. Head gesture recognition via dynamic time warping and threshold optimization. In: 2017 IEEE Conference on Cognitive and Computational Aspects of Situation Management (CogSIMA). Georgia, USA: IEEE; 2017

[12] Pierson A, Vasile C-I, Gandhi A, Schwarting W, Karaman S, Rus D. Dynamic risk density for autonomous navigation in cluttered environments without object detection. In: 2019 International Conference on Robotics and Automation (ICRA). Montreal, Canada: IEEE; 2019

[13] Sezer V, Dikilita C, Ercan Z, Heceoglu H, Oner A, Apak A, et al. Conversion of a conventional electric automobile into an unmanned ground vehicle (UGV). In: 2011 IEEE International Conference on Mechatronics. Istanbul, Turkey: IEEE; 2011

[14] Babu GA, Guruvayoorappan K, Variyar VS, Soman K. Design and fabrication of robotic systems: Converting a conventional car to a driverless car. In: 2017 International Conference on Advances in Computing, Communications and Informatics (ICACCI). Manipal, India: IEEE; 2017

[15] Quigley M, Conley K, Gerkey B, Faust J, Foote T, Leibs J, et al. ROS: An open-source robot operating system. ICRA Workshop on Open Source Software. 2009;**3**(3.2):5

[16] Abdelrasoul Y, Saman ABSH, Sebastian P. A quantitative study of tuning ROS gmapping parameters and their effect on performing indoor 2D SLAM. In: 2016 2nd IEEE International Symposium on Robotics and Manufacturing Automation (ROMA). Ipoh, Malaysia: IEEE; 2016

[17] Wang Z, Simoncelli E, Bovik A. Multiscale structural similarity for image quality assessment. In: The Thrity-Seventh Asilomar Conference on Signals, Systems & Computers. California, USA: IEEE; 2003

Chapter 6

Embedded Devices Security Based on ICMetric Technology

Khattab M. Ali Alheeti, Duaa Al_Dosary and Salah Sleibi Al-Rawi

Abstract

An intelligent wheelchair application is required which is equipped with the MEMSs which are magnetometer, gyroscope, and accelerometer sensors. The generated process of ICMetrics number is heavily based on magnetometer, gyroscope, and accelerometer sensors. In addition, this number can be utilised to provide the identification of device. Our proposed system passed through three phases. The first phase is bias reading that was extracted from MEMSs (gyroscope, magnetometer, and accelerometers) sensors; whereas, in the second phase, ICMetric number is generated by using the sensor bias readings that was extracted in the first phase. Therefore, this number is non-stored and can be utilised to provide identification of device. In the third phase, the security system is tested/evaluated to measure its effectivity. In other words, it is tested with dataset that was extracted from the trace file of ns-2. In this phase, performance metrics are calculated, which are rate of error, confused metrics, and accuracy.

Keywords: ICMetric technology, security, intelligent wheelchair, application, MEMSs, magnetometer, gyroscope, accelerometer

1. Introduction

The increasing number of embedded systems is subjected to a growing number of threats as community of the hacker is beginning to exhibit interestingly to these systems. Moreover, the achievement of security is not easy consequent to resources constraints of these devices. In this chapter, we demonstrate the basic concepts of security and explain the role of applying identification in the promotion and development of embedded device security using ICMetric technology.

This chapter presents the following concepts:

- Define the basic concepts of security and attacks.
- Explain embedded devices security.
- Describe ICMetric technology and its phases.
- Describe the type of application and sensors utilised.
- Explain statistical analysis and how ICMetric number can be generated for identification of embedded devices.

IntechOpen

- Explain how performance metrics of a system can be evaluated.
- Conclusion.
- Future works.
- References.

2. Security concept

Security aims for achieving the protection of system information in order to preserve the data from manipulation and theft and to attain the availability, confidentiality, and integrity. It includes protection of hardware, software, information, and telecommunications [1].

Security is a term used to describe different states, such as lack of risks and threats situation, prevention of risks, or achieve confidence. Achieving security is required in many areas at the level of individuals and organisations. Ensuring security requires individuals with competence and experience, so the level of security varies from one organisation to another. To ensure better security for organisations and individuals, it is important for network users to use the systematic approach, which involves analysing, designing, implementing, and maintaining a required network security system [2].

There are many goals of security as illustrated below:

- *Authentication*: means verifying the identity of a device or person.
- *Confidentiality*: means preserving the information confidential to prevent unauthorised access to the information. Confidentiality loss of information means disclosure of information which leads to loss of information.
- *Availability*: means the data is available when needed. Failure to access information in a timely manner causes a system malfunction.
- *Integrity*: means ensuring that the information is sound from manipulation and destruction. Lack of integration means that information is subject to modification or sabotage.
- *Non-repudiation*: Ensures that information can never deny ever sending or receiving the message.
- *Access control*: block illegitimate or unwanted access by restricting access only for authenticated entities.

In order to gain illegitimate access, adversaries are capable of exploiting system weaknesses. As systems move out from homes and offices security to more settings in every place, security importance cannot be denied. It is essential to guarantee the hardware and software security of any system.

3. Security attacks

In order to protect systems from attacks, security experts must assess and identify the risks and vulnerabilities of the system and define how to use mechanisms that ensure security for the safety of the system. Section below presents a discussion of possible attacks of the system and their propagation in daily life.

3.1 Attacks types

Many types of attacks are described below with some detail:

1. Focused attack: this type of attack focuses on specific systems and does not restrict to money, resources, and time. The most practical examples of these type of attacks are targeting to defence installations and penetrating enemy communication lines.

2. Cryptanalytic attacks: cryptanalysis attacks have the ability to decrypt the ciphered text without accessing the encryption keys. These attacks are combined plain and cypher text attacks.

3. Network attacks: at present, systems are vulnerable to external attack through networks. This type of attack is done through monitoring, password hiding, and spoofing.

3.2 Targets of attacks

The system is exposed to different types of attacks; some of these attacks are presented in the following subsections. Generally, expected attacks are divided into three groups: attacks against availability, authenticity, and confidentiality :

1. Threats on availability: attacks' types on availability are the strongest, and we will mention two categories which are:

    a. Message suppression attack: packets are dropped by the attackers from the network, they exploit these packets in other time. This type of attack causes a lot of problems in the network.

    b. Drop the package: black hole attack is a dropping attack where the existence of this type of attack is the reason of package losing.

2. Threats to integrity: an example of this threat is alteration attack that happens when the attack changes message content. Re-send or delay of the message is considered as one of the forms of this attack.

3. Threats on authentication: fabrication attack is an example of this threat where attackers can gain their goals by broadcasting false messages in the network. These false messages are certificates, warnings, and identities.

4. Embedded devices security

An embedded system is a special-purpose computer system, which is fully encapsulated in the device it controls. These systems have special requirements and complete pre-defined tasks [3]. It consists of a combination of hardware and software; it is designed for dedicated functions. Embedded systems include several elements; each of them has a pre-defined task, and these differ from the typical desktop or laptop computer.

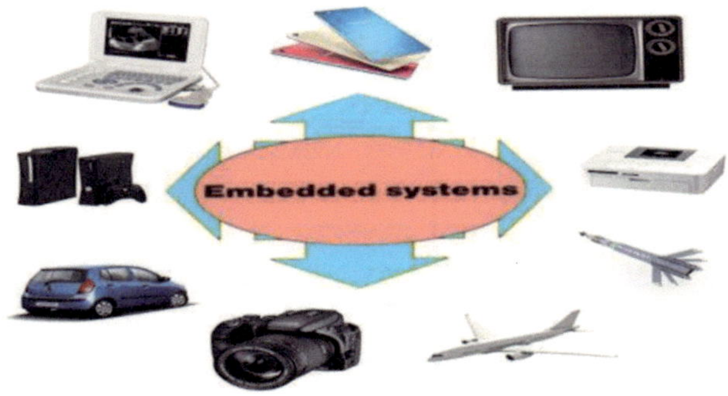

Figure 1.
Embedded system applications.

Embedded system spread in many devices, and the users of these devices are capable of performing almost all the network/internet applications that run on these devices. These devices are also involved in transport of secure data over public networks that require defence from unauthorised access.

As a result, security in embedded systems has spread every passing day in many fields like aerospace, telecom, healthcare, and wearable devices. **Figure 1** illustrates embedded systems applications such as railways, mobile phones, consumer electronics, tables, laptops, and healthcare application.

Embedded devices spread in a wide range of applications, and these devices handle critical information. For this reason, it is desirable to have some security mechanism deployed on embedded devices either in the form of software or hardware. However, embedded devices security is a challenging function and treated an open research case due to the resource-constrained nature of these devices. The security of embedded systems can become an issue, even bigger than the insufficiency of security of current desktop computers. The reason for this lack of security is hardware devices' constraints when performing measures of security and security cost. Manufacturers attempt to reduce costs of production to gain a market advantage for price critical products.

Many of application is heavily based on embedded systems that present in all our lives aspects. Embedded system devices are often networked via wireless communication links to accomplish advantageous tasks. However, the communication channel that is characterised by the wireless nature between the embedded devices makes them vulnerable to attacks and adversaries. Therefore, security of embedded systems is a main aspect of embedded systems design and is currently a major field of scientific research.

5. Problem definition

Many protection techniques have been proposed that try to provide security still do not interest on the most serious matter about who has access to the system. An alternate approach for providing the security of system from measurable properties of a target device is named ICMetric. In this chapter, ICMetric technology exploits the characteristic and behaviour of an embedded system to obtain a collection of properties and features, which aims to uniquely identify and secure an embedded system based on its own behavioural identity. The ICMetric technology allows a device to generate an identity which is used for authentication and a range of other cryptographic services.

In addition, some security techniques depend on the stored keys to enable secure information. Such techniques have a failure when the stored keys compromised the security of any data protected by these keys. Thus, it is important to employ security mechanisms to protect these systems from attackers. ICMetric technology has been designed as a method of deterring key theft by exploiting features of device to provide ICMetric number that will be utilised for identification of device. ICMetric number is not stored on the system and generated only when it is required by lightweight equation.

6. Objectives

In this chapter, the ICMetric technology is utilised to develop embedded devices security. This technology proposes using features of a device to generate ICMetric number used for device identification. The objectives behind using ICMetric were to overcome the problems related to accuracy rate, the failure to detect new attacks and the increased number of false alarms.

In this chapter, the objectives are as follows:

- Utilising ICMetric technology that depend on features that make each device different to generate a single and unique number called ICMetric number used for device identification.

- Demonstrating that it is possible to use the features of a device to create an identification that provides security of embedded devices. To achieve this, possible properties and features are investigated which can be used for the creation of an ICMetric of a device.

- Solving security problem presented in embedded device by applying device authentication.

- Proving that MEMS sensors can be used to improve security and apply identification of device. Bias reading generated by these sensors can be utilised to provide identification of device.

- Applying some statistical and mathematical analysis on bias readings extracted from MEMS sensors to provide ICMetric number used for identification and other security services.

- Training and testing dataset by using Support Vector Machine (SVM).

- Testing and evaluating system performance metrics, which are confused matrix, accuracy rate, and error rate.

- Evaluating additional performance metrics such as packet delivery rate, throughput rate, and end-to-end delay rate.

7. Integrated circuit metric technology

Encryption systems rely on the use of algorithms which in turn depend on the use of the secret key which is stored. Trying to increase the size of the key to stop the brute force, but increasing the size of the key cannot always protect the security of the system and deter theft [4].

In order to eliminate the theft and by relying on the special features of each device, we can create an identification for each device, and this identification is called ICMetric. Other hardware techniques differ from the ICMetric technology in the selection of device characteristics. Traditional fingerprinting techniques depend on the characteristics that are easily exposed to capture, deception, or repetition by the attackers. ICMetric technology uses internal behaviour that increase the complexity of generated ICMetric and they are hard for an attacker to predict or spoof at runtime such as features that can be employed for creation an ICMetric are Media Access Control (MAC) addresses and serial numbers. Other features are utilised, which are application usage special task, such as browsing histories, camera resolutions, common user files, and system profiles [4].

ICMetric technology retains the idea of storing the key where there is no encryption key in the system; this will reduce the attackers. It uses hardware and software features of device to create ICMetric, which will be used in encryption services. There is a similarity between the biometric systems and the ICMetric, as these systems used features for identification of different persons. Similarly, ICMetric uses the characteristics of the device to identify each device uniquely and thus eliminates the idea of stored keys and deters theft of stored keys.

ICMetrics points to a new technique that can be employed to extract features from the hardware and software environment of a system. Every device is singular in its internal environment and then the features that make every device diverse can be employed to create a unique and single number for each device. It is based on the next concepts:

1. ICMetric number is not stored on the system and can be recreated when needed.

2. If the system is attacked, there will be no theft because the ICMetric is non-stored.

3. ICMetric number and any proceeding outcomes that are based on the ICMetric number will change if any adjusting has been done with the software, hardware, or environment.

4. There is no requirement to store any template that can serve the aim of device validating.

The generation of ICMetric system is comprised of two phases: calibration phase to collect detailed knowledge and operational phase to distribute each extracted features for typical sensors.

7.1 Calibration phase

At this phase, the characteristics are documented and analysed, normalisation distributions are utilised on feature values noticed in the system. A device ICMetric basis number can be created by applying statistical and mathematical operations on the extracted feature values. This phase is utilised once only when the system needs the ICMetric basis number. The features on which the ICMetric is based are unique; therefore, it is difficult for the attacker to detect or generate it. This is an important case to improve the ICMetric strength.

7.2 Operational phase

Operational phase follows the calibration phase where the unique number is generated depending on the extracted features. The preprocessing phase can be applied to generate unique features that distinguish it from others [5]. In this phase, an effort is made to generate a resulting device ICMetric basis number through either feature concatenation or feature addition.

8. Intelligent wheelchair application

Some companies produce different types of an intelligent wheelchair. An intelligent wheelchair can be defined as a uniquely modified powered wheelchair which is provided with a control system and variant sensors. Intelligent wheelchair is designed to provide several services to users in different ways. It eliminates the user's responsibility for moving the wheelchair. User types of intelligent wheelchairs are different according to their situations and disabilities. According to this, there will be different designs of intelligent wheelchair. The aim of intelligent wheelchair is to grant higher independence to people with lower mobility such as disabled or elderly individuals [6]. **Figure 2** states intelligent wheelchair.

In this chapter, ICMetric technology is integrated into an intelligent wheelchair. MEMS sensors embedded in intelligent wheelchair are utilised to ensure effective usage and to provide identification of intelligent wheelchair.

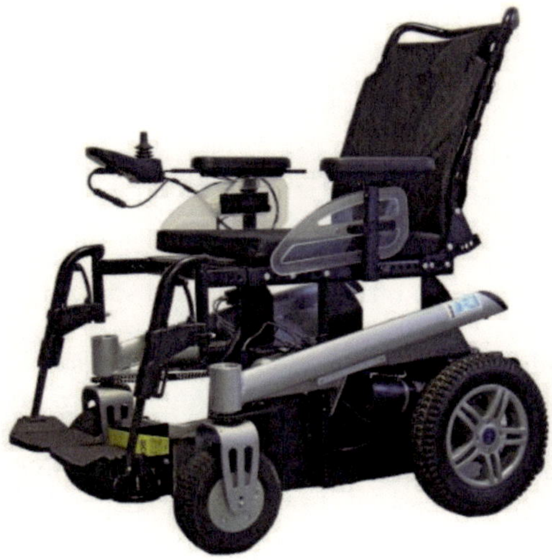

Figure 2.
Intelligent wheelchair.

9. Appling ICMetric technology in intelligent wheelchairs

In this chapter, ICMetric technology is integrated into an intelligent wheelchair. In order to ensure effective usage of the intelligent wheelchair and the need to protect the safety of each wheelchair, it is advantageous to provide identification of intelligent wheelchair to confirm the user's right to access the system and information and defend against identity theft and fraud. For achieving these aims, ICMetrics represents a new method for generating unique identifiers for embedded devices and improves security by reducing fraudulent activity.

ICMetric technology can improve secure communication between devices, reduce fraudulent activity, prevent unauthorised access to the systems and devices connected with the wheelchair, implicit detection of tampering of the software or hardware associated with the wheelchair, and prevent the fraudulent cloning or imitation of the electronics associated with the wheelchair.

While many security techniques are now developed, these cannot necessarily defend against unauthorised activity when the security and safety cannot be absolutely assured. The use of ICMetric technology to provide identification represents a new concept of controlling access to devices and is explicitly aimed at providing protection against attacks and improving security.

The ICMetric security system proposed in this chapter uses bias readings that have been extracted from sensor devices. These readings exploited to create ICMetric basis numbers that were working as identification for device.

In this chapter, suitable features can be gotten from the sensors to describe behaviour of intelligent wheelchair such as the gyroscope, magnetometer, and accelerometers. The offset is utilised in the sensor measurement to propose a security system that apply an ICMetric basis number using the sensor bias readings.

10. Intelligent wheelchair sensors

In order to apply features measuring of wheelchair, specific hardware circuits along with a software-based monitoring infrastructure are needed to be prepared and integrated in the system. Features can be read from various integrated sensors, and large origin of features is also the system's behaviour. For obstacles avoiding, intelligent wheelchairs need sensors to perceive their surroundings. Many sensors are used by intelligent wheelchairs as explained below:

- Ultrasonic sensors (i.e. sonar) and sonar sensors are very precise when the sound wave emitted by the sensor strikes an object at a right angle or head on.

- Infrared (IR) sensors emit light, rather than sound, and can be fooled by dark or light absorbent material rather than sound absorbent material.

- Laser Range Finders (LRFs) offer a 180°, two-dimensional scan within the plane of the obstacles in the environment. Another option is a 'laser striper', which contains a laser emitter and a charge-coupled device camera. The image of the laser stripe returned by the camera can be used to determine distances to obstacles and drop-offs based on breaks in the stripe.

In this chapter, MEMS sensors, which are magnetometer, accelerometer, and gyroscope, are used. MEMS sensors have many applications in measuring either acceleration or angular velocity about one or several axes as an input to control a system. Some details of accelerometer, gyroscope, and magnetometer sensors are presented below:

10.1 MEMS accelerometer in intelligent wheelchair

The MEMS accelerometer is a highly sensitive sensor and capable of detecting the tilt. This sensor changes the direction of the wheelchair depending on tilt. For example, if the tilt is to the right side, then the wheelchair moves in the right direction, and if the tilt is to the left side, then the wheel moves in the left direction. The wheelchair movement can be controlled in forward, reverse, left, and right direction with obstacle detection using ultrasonic sensor. This wheelchair automatically senses the presence of an obstacle in its path and turns its direction of movement.

10.2 MEMS gyroscope in intelligent wheelchair

The MEMS gyroscope sensor provides an angular velocity of the wheelchair's wheel as compared to translating the acceleration values to rotation angles for calculating heel rotations. Angular velocities are used directly to estimate linear speeds and distances travelled, which allow us to provide wheelchair users with real-time feedback through smartphone applications.

10.3 MEMS magnetometer in intelligent wheelchair

Intelligent wheelchairs used MEMS magnetometer sensors for measuring and detecting magnetic fields. Hall effect, magneto-resistive effect, and fluxgate effect are the most popular principles in magnetometer sensors. Magnetometer sensor measures the magnetic fields based on Hall effect.

ICMetric technology presented here uses bias readings that have been generated from sensor devices. These readings are used to apply ICMetric basis numbers that were utilised as identification for device. Sensor-based identification field has proved that the use of sensory data is possible and that it is feasible to provide device identification.

11. The ICMetric security system

Current defensive mechanisms are not enough for preventing the internal attacks in device, since they require ICMetric security system as protection system to increase their security. ICMetric technology depends on measureable features, which have been achieved from the properties of a particular embedded system. Features are generated in the particular system that represents a unique feature for that system. The focus is on utilising a magnetometer, gyroscope, and accelerometer sensors that are provided in the new system.

In this chapter, an intelligent wheelchair is required where the bias readings extracted from sensors embedded in intelligent wheelchair are utilised in the ICMetric security system generation. These bias readings were employed to create an ICMetric basis that was used as identifications for device. In this chapter, ICMetric security system based on bias readings is extracted from gyroscope, accelerometer, and magnetometer sensors. The proposed algorithm is summarised in the following steps:

Algorithm (3.1): The Implement ICMetric
Input: Behaviour features that extracted from trace file of ns-2.
Output: Normal behaviour or abnormal behaviour.
Step 1: Establishing parameters for simulation. Step 2: Generating mobility and traffic model. Step 3: Extracting features from trace file. Step 4: Pre-processing of the extracting features. Step 5: Integrating of ICMetric number generated by MEMS sensors. Step 6: Dividing the extracted dataset into three groups which are training set, testing set, and validation set. Step 7: Training phase. Step 8: Testing phase. Step 9: Results.

12. Sensors bias measurement

In some cases, it is not feasible to collect the bias in the sensor. Sensors are required for generating the bias readings of a system which do not require user intervention and are not influenced by external factors. MEMS is a technology that combine mechanical and electrical components. There are many MEMS-based sensors that are being embedded into recent vehicles, wearable devices, laptops, and smartphones. The most commonly used and good examples of MEMS sensors are the accelerometer, the magnetometer, and the gyroscope. Accelerometers are intended to measure the acceleration of an object, whilst gyroscopes measure angular velocity. Magnetometer sensors are used for measuring and detecting magnetic fields.

MEMS sensors are used in this chapter because they are readily available and also the required stimulus is easy to create. Various numbers of bias reading are obtained from MEMS sensors to generate ICMetric number. The system needs to determine the optimal number of readings which are used in identification processes to control the stability of the statistical processes of the ICMetric generation. We need to calculate the population mean and compare the result with the mean value calculated for a smaller subset of readings to determine the best number of readings.

In order to extract reading from sensors, MEMS sensors are placed on the board, simulation is applied to produce a constant bias. The case is similar when a sensor is under operation where accuracy of sensor output is affected by unclear damages due to mistreating. For bias generation, the stimulus must be specified. One of the advantages of stimulus is that it does not need a specific device to evaluate it but is equipped by the user. The readings generated by the sensor must be equal to the stimulus applied to the sensor. Every axis owns a different bias, which is showed in the readings. Experiments prove that the bias in every sensor is unique and reproducible. These bias readings are utilised to provide identification of device and improve security.

13. Statistical analysis for ICMetrics

ICMetric is not stored but is created when needed this distinguishes the technology of ICMetric in protection from attacks. Since it is created when needed, it requires simple mathematical processes and a statistical analysis for the values of features. Below are some statistical analyses required for the generation process of the ICMetric number is utilising to apply identification of device.

If we assume that $\overline{X}$ represents mean, x represents particular sample reading from accelerometer, magnetometer, and gyroscope, and n is total number of reading, then [7]:

$$\overline{X} = \frac{1}{n}\sum_{i=1}^{n} x_i \tag{1}$$

In order to complete ICMetric generation process, we need to calculate σ^2 as explained below, where σ^2 is the standard deviation.

$$\sigma^2 = \sum_{i=1}^{n} p(x_i)\,(x_i - \overline{X})^2 \tag{2}$$

Furthermore, other statistical and mathematical functions are utilised to analyse the generated reading. For example, we need to calculate variance (s^2) as explained in the following equation:

$$s^2 = \frac{1}{n-1}\sum_{i=1}^{n} (x - \overline{X})^2 \tag{3}$$

where s^2 is a measure of dispersion for extracting readings.

The skewness distribution (S) is a measure of asymmetry of the probability distribution of bias readings. It can be negative or positive:

$$S = \frac{3\,(\overline{X} - m)}{s^2} \tag{4}$$

To prove the uniqueness of the bias generated from accelerometer, magnetometer, and gyroscope sensors, we use 95% confidence interval. If $\overline{x}$ is the mean, σ is the standard deviation, and n is the total number of observations, then the confidence interval CI is given in the Eq. (5) where the numeric value v here equals to 1.96.

$$CI = \overline{X} \pm v \frac{\sigma}{\sqrt{n}} \tag{5}$$

In addition, other statistical and mathematical functions are utilised to analyse the generated reading such as inter quartile range (IQR) that represents difference between the third and the first quartile in offset data. IQR can be calculated according to the Eq. (6), where $Q3$ represents upper quartile and $Q1$ represents lower quartile.

$$IQR = Q3 - Q1 \tag{6}$$

14. Performance metrics

ICMetric security system has been evaluated using a number of ways based on trace file evaluation. The ICMetric security system can be generally evaluated from two views [8]:

1. Accuracy: this portion also termed effectiveness classification characterises the ability of the system to separate between intrusive and non-intrusive activities.

2. Efficiency: this portion deals with the resources required to be allocated to the system including CPU cycles and main memory

System features can be evaluated in terms of performance, correctness, and usability. Researchers used metrics to assess the performance of the system. Many performance measures are used to evaluate the system, which are based on the dataset extracted from the trace file created by ns-2.

The identification rate and four alarms are utilised as performance metrics to test the system. To measure and evaluate the system performance, four types of alarms are needed to calculate: true positive (TP), false positive (FP), true negative (TN), and false negative (FN). The measures will be calculated as follows [5]: Let

$$TP = \text{normal connection record classified as normal}$$

$$\text{TN} = \text{attack connection record classified as attack}$$

$$\text{FP} = \text{normal connection record classified as attack}$$

FN = attack connection record classified as normal

Then,

$$TP_{Rate(\text{sensitivity})} = \frac{TP}{TP + FN} \tag{7}$$

$$TN_{Rate(\text{specificity})} = \frac{TN}{TN + FP} \tag{8}$$

$$FN_{Rate(1\text{-sensitivity})} = \frac{FN}{FN + TP} \tag{9}$$

$$FP_{Rate(1\text{-specificity})} = \frac{FP}{FP + TN} \tag{10}$$

In addition, some extra metrics are utilised to evaluate system performance such as packet delivery rate (PDR), throughput, and end-to-end delay.

- Packet Delivery Ratio (PDR): the ratio between the number of packets sent from the origin and the proportion of packets received at the destination.

$$\text{PDR} = \Sigma N_r / \Sigma N_s \tag{11}$$

where N_r = number of packets received and N_s = number of packets sent.

- Throughput: the total number of packets that are transferred in the system. Throughput of a system can be presented as shown in the following equation.

$$\text{Rate of throughput(kbps)} = N_r * S/ST\ (12) \tag{12}$$

where N_r = number of packets received and S = packet size and ST = simulation time.

- Average end-to-end delay: the average time for packets reaching from the origin to the destination. Average end-to-end delay is explained in the following equation:

$$\text{Rate end - to - end delay (ms)} = \left(\frac{\Sigma end_{time} - start_{time}}{\Sigma N}\right) \tag{13}$$

where N represents a number of connections.

15. Conclusions

The main contribution of the present chapter is to achieve an identification model for intelligent wheelchair application with high rate of accuracy and with low error rate. This was done through the design of an identification process by using ICMetric technology. From the given results, the following substantial remarks were obtained:

1. ICMetric technology can be used to apply identification and improve security of embedded devices.

2. ICMetric technology relies on the special internal features of device where each device is unique in its internal environment.

3. MEMS gyroscope, magnetometer, and accelerometer sensors embedded in intelligent wheelchair are utilised. Three axes readings achieved from every sensor where every sensor will have unique readings.

4. Readings generated from MEMS sensors are analysed statistically to generate ICMetric number utilised for device identification. A statistical study of the readings generated from sensor shows practical use of MEMS sensors for the generation of a device identification.

5. Support vector machine exploited in this chapter to evaluate and test system under certain conditions.

6. In order to evaluate system performance, we need to calculate the performance metrics, which are accuracy rate, error rate, and four types of alarms.

7. Additional performance metrics can be evaluated for system such as, PDR, throughput, end-to-end delay.

During the training and testing processes, the SVM with dataset is extracted from the trace file generated by ns-2; it was found that using ICMetric technology for embedded devices identification provides better rate of accuracy and low error rate.

16. Future work

1. This chapter has established the design of an ICMetric by using different features of device. Therefore, the aim of the future research is to discover more features which can strengthen the device ICMetric.

2. The ICMetric technology has not been researched in bitcoins and block chains. It can be integrated into bitcoins to provide secrecy of transactions.

Author details

Khattab M. Ali Alheeti[1*], Duaa Al_Dosary[2] and Salah Sleibi Al-Rawi[3]

1 Computer Networking Systems Department, College of Computer Sciences and Information Technology, University of Anbar, Iraq

2 Computer Sciences Department, College of Computer Sciences and Information Technology, University of Anbar, Iraq

3 Information Systems Department, College of Computer Sciences and Information Technology, University of Anbar, Iraq

*Address all correspondence to: co.khattab.alheeti@uoanbar.edu.iq

References

[1] Stallings W. Cryptography and Network Security: Principles and Practice. 5th ed. Prentice Hall; 2011

[2] Alshahrani M, Teymourlouei H. Network security: Threats and vulnerabilities. In: International Conference on Security and Management. CSREA Press; 2016. pp. 115-121

[3] Lizarraga J, et al. Security in embedded systems. In: IADIS International Conference Applied Computing; 2006. pp. 697-699

[4] Tahir H, Mcdonald-maier K. Securing health sensing using integrated circuit metric. Sensors. 2015;**2015**(15):26621-26642

[5] Ali KM. Intrusion detection system in external communication for self-driving vehicles [PHD thesis]. School of Computer Science and Electronic Engineering, University of Essex; 2017

[6] Faria B, Mónica S, Vasconcelos L, Reis P, Lau N. A methodology for creating intelligent wheelchair users' profiles. In: ICAART 2012—Proceedings of the 4th International Conference on Agents and Artificial Intelligence. 2012 SCITEPRESS (Science and Technology Publications). Vol. 1. 2012. pp. 171-179

[7] Tahir H, Tahir R, Mcdonald-maier K. Securing MEMS based sensor nodes in the internet of things. In: 2015 Sixth International Conference on Emerging Security Technologies Securing; 2015. pp. 44-49

[8] Kumar G. Evaluation metrics for intrusion detection systems—A study. International Journal of Computer Science and Mobile Applications. 2014;**2**:11-17

Chapter 7

Manipulating Complex Robot Behavior for Autonomous and Continuous Operations

Chengliang Liu, Liang Gong and Wei Zhang

Abstract

Service robot control faces challenges of dynamic environment and complex behavior, which mainly include eye-hand coordination and continuous operations. However, current programming scheme lacks the ability of managing such tasks. In this chapter, we propose a methodology of software development paradigm for the continuous operation of the dual-arm picking robot. First, a dual-arm robot is built for picking with the purpose of selectively harvesting in plant factory. Second, a hierarchical control software is framed by means of "Sense Plan Act" (SPA) paradigm. Third, based on the previous design, programming concept, and the ROS system, the sub-node programming of visual module, motion module, eye-hand coordination module, and task planning module are implemented with a state machine-based architecture. The experimental results show that if total number of targets within the visual field is not more than three, the average picking time is less than 35 s. The fluency of concurrent task management shows the feasibility of manipulating complex robot behavior for autonomous and continuous operations with the finite state machine model and task level architecture.

Keywords: dual-arm robot, complex behavior, continuous operation, robot operating system (ROS), finite state machine

1. Introduction

With the development of technologies such as industrial robots and computer image processing, a series of research and experiments on intelligent picking robots have been carried out in Japan and other related countries, such as tomato, apple, and grape picking robots [1]. Research on picking robots has focused on two parts: the first is a hardware device that can achieve rapid picking, that is, how to design a stable, efficient, and adaptable mechanical and visual sensing system, and the second is to design intelligent software for continuous operations, that is, to accurately identify, distinguish, and locate targets and to solve the problem of task planning and task scheduling during continuous picking operations.

In the research field of picking robots, Kondo et al. [2] designed a tomato picking robot using a 7-degree-of-freedom manipulator. It used fingers and pneumatic nozzles in conjunction with a color camera to complete the picking operation. The experiment achieved a picking success rate of about 70% [3]. Tanigaki et al. [4]

IntechOpen

developed a cherry picking robot using a four-degree-of-freedom manipulator and a specially designed end effector with suction and shear functions. The visual part uses a light emitter, a photodetector, and a scanning device. The fruit was picked in 14 s, and the success rate was about 84%. The CROPS plan completed in 2014 was jointly completed by many European countries and units and aims to develop a modular picking robot system for different mission scenarios. The greenhouse bell pepper picking robot platform completed in the experiment in the Netherlands [5] used a 9-degree-of-freedom manipulator, two color CCD cameras and a depth-measuring camera, and its end effector was also equipped with a small camera to complete the picking with higher accuracy. Taqi et al. [6] have developed small household cherry tomato picking robots that can achieve very accurate picking tasks in specific environments.

The continuous operation of the picking robot can make intelligent decisions on multi-task under multi-objective scenarios and plan the operation according to the picking needs. The research goal is that the picking robot system can intelligently select and pick fruits that meet the picking conditions, thereby greatly improving the degree of automation of the picking process and improving the quality of the harvested fruits. Because the picking robot is still in the laboratory research stage, the research on continuous operation is also very limited, and it is basically in its infancy. Japan's Nagata et al. [7] used the shape to judge and classify strawberry quality. The accuracy in the experiment is acceptable, but the speed is much slower than artificial. Zhao et al. [8] investigated the visual recognition of apple maturity, using multi-spectral laser beams to complete fruit identification and positioning and ripeness judgment. Guo [9] and others judged strawberry maturity based on HIS color space algorithm. In the field test, the accuracy of the goal of picking fruits with maturity of 80% or more was more than 90%. Wang [10] and Ling et al. [11] carried out research on selective harvest information acquisition and path planning of tomato picking robots. Multi-sensor information fusion method was used for tomato quality detection and classification and selective picking decision. The appearance maturity of the fruit is detected by the H-means in the computer image, the fruits are classified in real time according to the agricultural industry standards, and the selective harvesting decision is made through the progressive identification of feature information and the fusion decision. At harvest time, path planning is performed on multiple targets through an optimization algorithm and then boxed by level.

This chapter mainly focuses on developing an intelligent software system for the continuous operation of the dual-arm picking robot in a plant factory. Typically, the semi-structure environment of the greenhouse poses challenges for autonomous operations of the robot, and the complex tasks mainly include identification and positioning under variable light conditions, selective picking in multi-cluster growth environment, and complex multi-task programming. For the difficulty of developing the software system, a hierarchical modular software system framework is designed. Moreover, a scheduling method of functional modules is designed based on the idea of finite state machine for the complex multi-task planning problem of tomato continuous picking process. The task scheduling design based on the Finite State Machine (FSM) reduces the difficulty of development work and improves the efficiency of development.

This chapter is organized as follows: Section 2 briefly describes the hardware structure of a dual-arm robot, and Section 3 presents the software framework with the highlights of deploying SMACH- and ViSP-based nodes in a ROS development environment. The experimental results are also included at the end of Section 3.

2. Hardware design of a dual-arm robot

The semi-structured operating environment in the plant factory is relatively complicated, such as the occlusion of fruit branches and leaves, the challenging grasping shape of the fruit, the changing light, and the variety of tasks, which requires to consider the multi-tasking ability of the robot when designing the hardware of the picking robot. At the same time, the complexity of the picking environment requires that the execution of the robot be robust to the environment in which the target is located, only in this way can it pick fruits in different states. Based on the above two design goals, we designed a dual-arm picking robot to simulate human picking operations.

2.1 Design of dual-arm robot body

The mechanical structure of the robot body mimics human arms, and the left and right arms each have three joints: a vertical lift joint, a boom rotation joint, and a forearm rotation joint. The vertical lifting joint is driven by a servo motor to drive the roller screw. The actual effective stroke is about 300 mm. The big and small arm joints are driven by a servo motor connected to a harmonic reducer. There is a waist joint between the body and the base, which can provide 360° rotary motion. The single-arm movement of the robot is similar to that of the SCARA robot. The vertical positioning is achieved by the lifting joint, and the rotation of the large and small arms realizes the positioning in the plane. There are three degrees of freedom in motion. Each of the left and right forearms is designed with a mounting flange surface, and end effectors can be installed as required. The dual-arm robot base is installed on a mobile cart and is transported along the track to different picking points for picking operations.

Due to the limitation of the freedom of the robot body, for complex tomato picking environments, it is not enough to rely only on the freedom of the arms, so we can use the design of the end effector to increase the freedom of our robot and enable the robot to flexibly complete the picking operation. We designed a shearing end effector on the cutting hand of the dual-arm robot and designed a suction-type end sleeve on the auxiliary hand to fix the target tomato and assist the hand in picking (**Figure 1**).

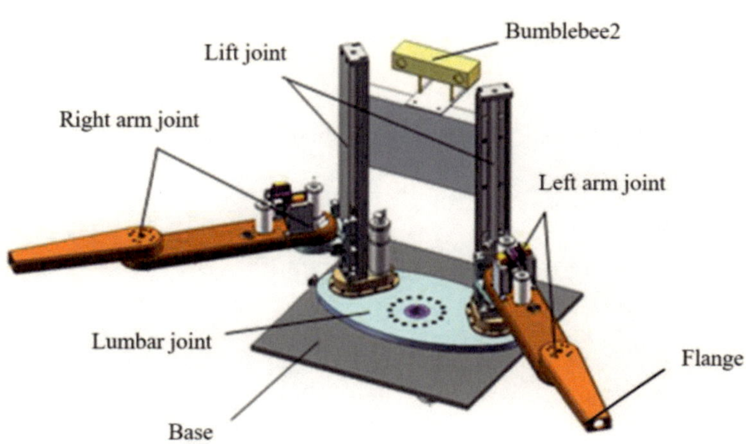

Figure 1.
The structure of the robot body.

2.2 Dual-arm robot coordinate system

As shown in **Figure 2**, the world coordinate system (x_w, y_w, z_w) of the dual-arm robot is built on the waist and coincides with the waist joint coordinate system. The left and right arm coordinate systems (x_l, y_l, z_l) and (x_r, y_r, z_r) use the right-hand principle, and the span direction at the zero position of the x axis is different from the x axis of the world coordinate system by +45° and −45°. The z-axis direction is vertically upward and is at the zero position at the lowest point. The x–y plane is parallel to the world coordinate system x–y plane. The binocular camera coordinate system is established at the intersection of the right-eye visual axis and the camera lens. The coordinate system adopted by the bumblebee2 camera is the left-hand principle. In order to be compatible with the entire system, the y-axis direction is reversed.

We use the tf function package [12] provided by ROS to maintain the transformation relationship between coordinate systems. After the robot model is built according to urdf, the system will automatically broadcast the transformation relationships between all the coordinate systems. The picking robot coordinate system is shown in **Figure 3**.

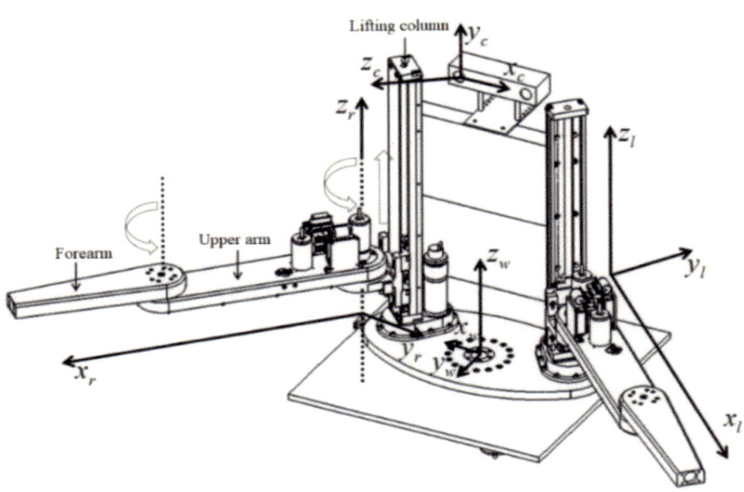

Figure 2.
Robot coordinate system.

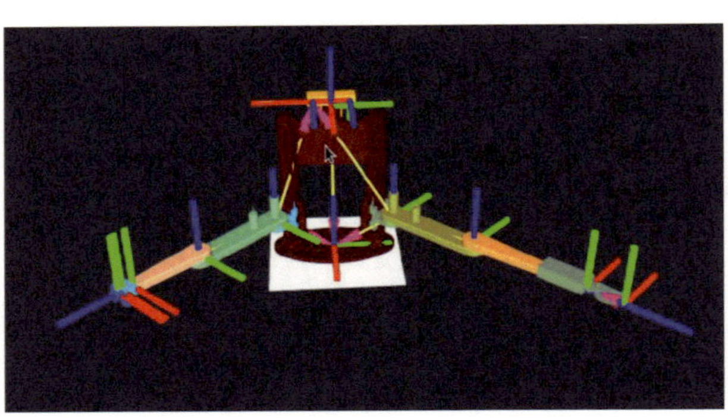

Figure 3.
The picking robot tf coordinate system under Rivz.

Based on the above coordinate system, using the *rosrun tf view_frames* command, we can view the tf tree of the dual-arm picking robot, as shown in **Figure 4**.

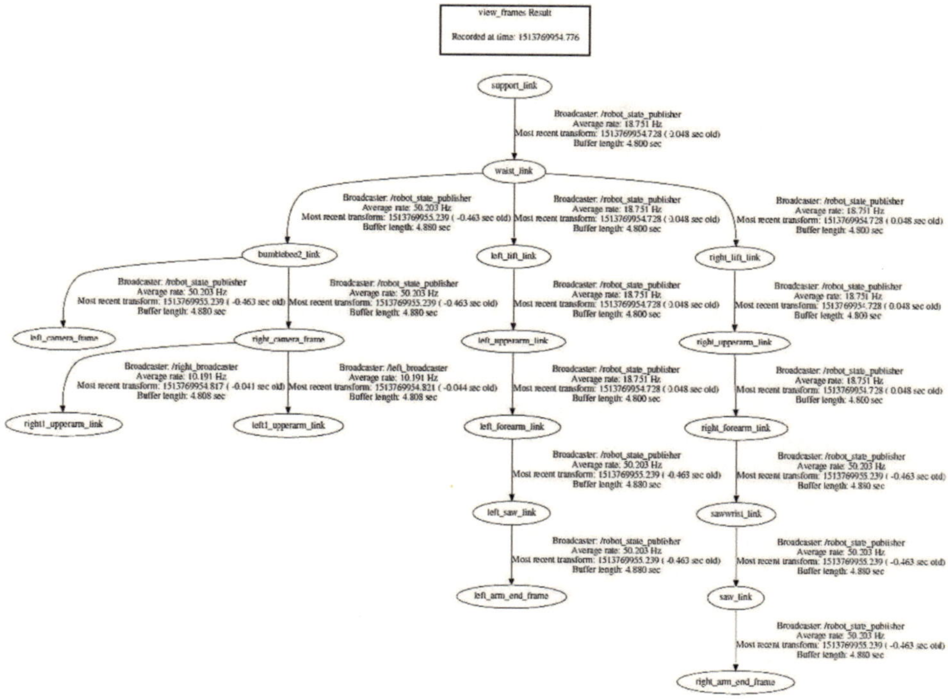

Figure 4.
The tf tree of robot.

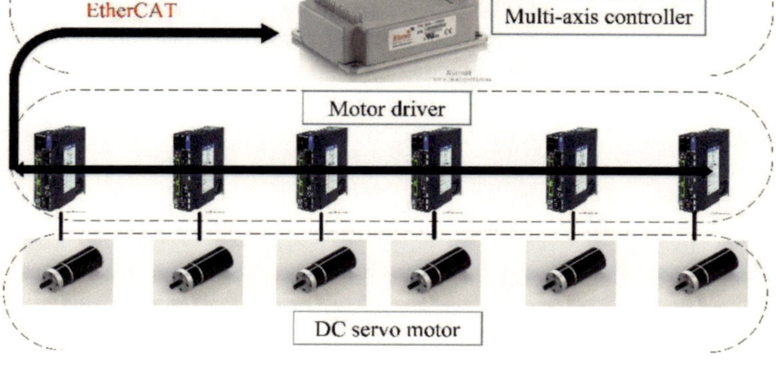

Figure 5.
Picking robot communication architecture.

2.3 Communication architecture of the dual-arm picking robot

The communication architecture of the dual-arm picking robot is shown in **Figure 5**. The motor driver is a Gold series motor driver produced by Elmo and is equipped with the same series of multi-axis controllers. The manufacturer has provided a complete motor driver to multi-axis controller communication protocol and communication protocol implementation and does not require customers to conduct secondary development. The multi-axis controller uses Modbus TCP communication as the communication between the lower computer and the industrial computer. It is connected to it through an Ethernet cable.

3. Software frame design

In the process of programming, we generally manually divide the tasks of the robot. Once the tasks are effectively divided, we can stack the tasks with the smallest functional components, so the difficulty of task division is how to use human prior knowledge to divide the minimum granularity of robot skills, which is a learning process. At the same time, how to easily and effectively combine the divided components is also an important part of completing the task simulation.

The system architecture describes the functional structure of the subdivision and the topological relationship between them and a series of specifications that need to be set for subsequent development. The basic requirements of software engineering include modularity, code reuse, and function sharing. Using a common framework is helpful for decomposing development tasks and code migration. Robot software also follows the general rules of software engineering. Architecture is how you break up the robot's functions and organize your code. A clear architecture that matches the project directly determines your development efficiency and even the success or failure of the final function. There are two main approaches to robot system architecture: SPA architecture and behavior-based architecture.

3.1 SPA

The software system architecture is "Sense Plan Act"(SPA), as shown in **Figure 6**. The robot maps the external environment space through sensors and uses a certain modeling method to structure and model the perception information and then analyzes the model to plan the robot's actions. Finally, the action instructions are executed in the environment to achieve a complete interactive process.

The typical software architecture in a SPA robot system is a three-layer architecture: the perception layer, the planning layer, and the motion control layer, as shown in **Figure 7**. The perception layer receives and processes the sensor data, the planning layer plans the motion trajectory, and the motion control layer ensures the accurate execution of the movement. The SPA robot software system architecture

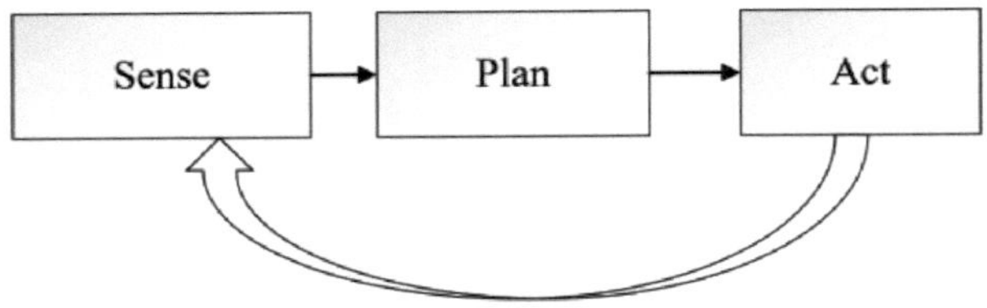

Figure 6.
SPA work pattern.

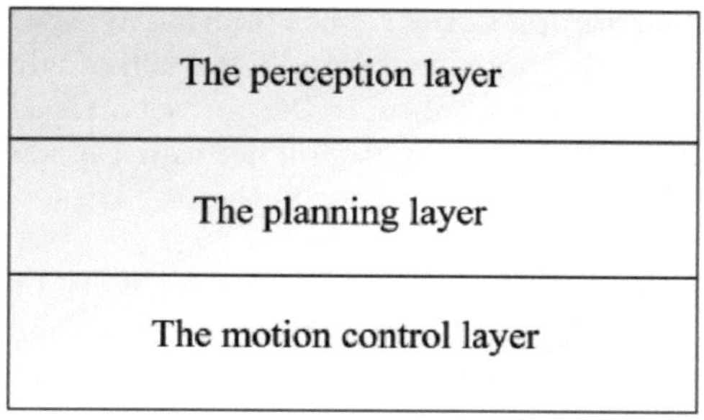

Figure 7.
SPA layered design.

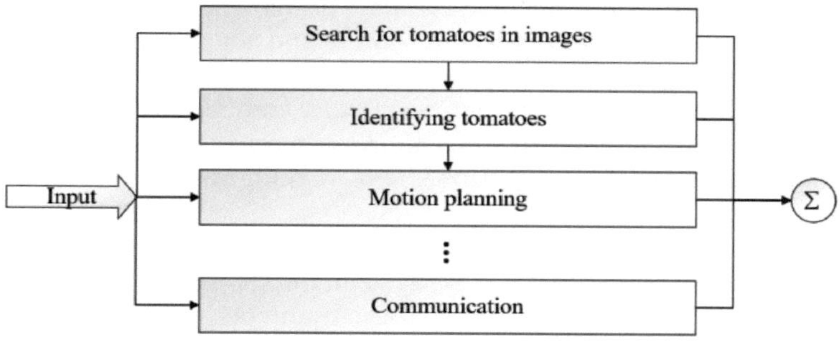

Figure 8.
Action-based software system architecture.

pays more attention to the perception and modeling of the world because this is the basis for the accuracy of subsequent planning and movement.

3.2 Behavior-based architecture

The behavior-based software system architecture is a top-down software design. The small functions of each robot are packaged into individual small modules. All functional modules can be executed in parallel without prioritization. A robot task can be understood as an organic composition of functional modules (**Figure 8**).

To a certain extent, all robot actions are responses to stimuli (inputs). This stress mode avoids the thinking logic in the SPA architecture and facilitates the rapid action response. In order to achieve the task, we can design a control scheme to change the stress level of the action. Therefore, we need a global controller to coordinate the choice of actions in order to achieve our intended purpose. The behavior-based software design framework has good flexibility, but it increases the difficulty of control. When multiple actions can affect the output, problems are easy to occur.

Therefore, combined with the SPA software architecture and behavior-based software architecture, we design a software framework that combines the advantages of both architectures for continuous operation of a dual-arm picking robot. Its characteristics are as follows:

1. Hierarchical modular design: The software architecture absorbs the advantages of the SPA architecture and also adopts a hierarchical design. The layered design is mainly logical, which makes it easier for users to understand the working mode of the robot. At the same time, it also absorbs the advantages of behavior-based architecture, that is, functional modularity. Based on the

analysis and understanding of the robot's internal architecture, we divide the functions of the robot into seven modules, each of which is functionally independent of each other. Combining the above two is a hierarchical modular design, which divides the functional modules into a specific layer according to the attributes of the functions, thus strengthening the logic of the system.

2. Finite state machine control: In order to solve the shortcomings of behavior-based architecture, we have designed a task planning module based on finite state machines, which is used to schedule and control the execution order of each functional module to complete a specific task.

3.3 Hierarchical modular design

Layering is an important concept in software design. The division of layers provides a framework for business decomposition and simplifies many thinking processes. Considering the design characteristics of the software system and the functional features of the robot, the entire software architecture can be divided into four layers: presentation layer, application layer, sense layer, and data layer. The hierarchical modular design architecture is shown in **Figure 9**.

1. Presentation layer: Presentation layer has more business logic requirements. We designed the presentation layer based on the QT architecture. The entire presentation layer includes several main components as shown in **Figure 10**: the RVIZ module displays the model of the robot and other visual information; the image module displays the video image information collected by the current robot; the node module monitors all current node information; and the DashBoard module provides users with a function module for manually operating the robot; shell module provides command line functions; console module displays all log information executed by the system; reconfigure module provides users with a convenient tool for changing model parameters; and diagnostic module provides real-time robot monitoring information.

2. Application layer: Application layer focuses on the task execution of a single robot. It is separated from the implementation of specific functions and uses a combination of function modules to coordinate a task.

3. Sense layer: Sense layer is responsible for the interaction between the software system and the hardware. There are both a visual module responsible for

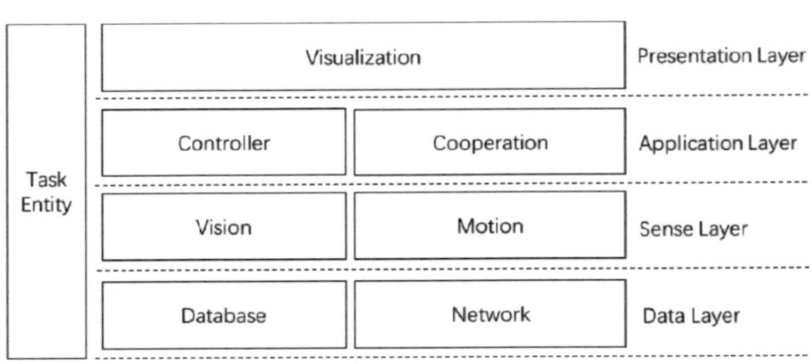

Figure 9.
Layered modular architecture design.

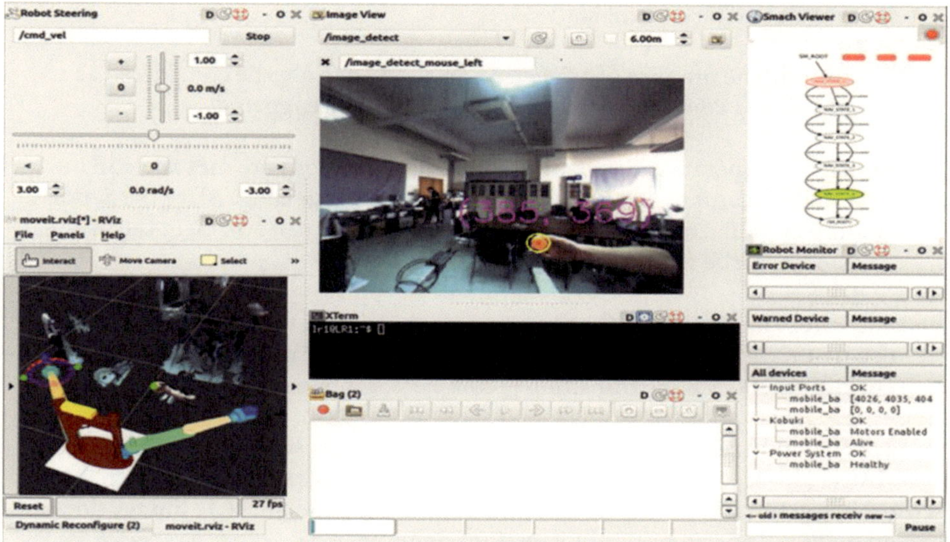

Figure 10.
Presentation layer.

environmental perception and a motion module responsible for the motion control of the upper and lower computers. Sense layer is a description of the robot's capabilities.

4. Data layer: Data layer serves the data generated by the system. Part of the data generated by the system is stored in a local database for real-time decision making of the system. One part is uploaded to the server through the network and is fused with data from other robots and other time dimensions to plan the continuous operation of the robot.

The criteria for the division of functional modules are to reduce coupling, relatively independent functions, and high code repeatability. According to the robot's task module, the software system can be divided into motion (*motion_pkg*), control (*control_pkg*), vision (*vision_pkg*), visualization (*visualization_pkg*), collaboration (*coorperation_pkg*), database (*database_pkg*), and network (*network_pkg*). Each function module is represented as a function package at the file system layer. There can be multiple nodes in a package, and different nodes can be written in different programming languages.

3.4 Cooperative control of dual-arm picking based on FSM

3.4.1 SMACH

Finite state machine (FSM) is a mathematical model of computational science. The objects it represents can be in a limited number of states at any one time. When an external input occurs, the system responds to the external input, and the FSM can conditionally transition from one state to another. This process is called an action. In computer science, finite state machines are widely used for modeling application behavior, hardware circuit system design, software engineering, compilers, network protocols, and computation and language research. FSM can be defined by the present state, condition, action, and substate. The specific interpretation is as follows:

1. Present state: The current state.

2. Condition: The premise of triggering an action can also be considered as an event. When a condition is filled, an action will be triggered.

3. Action: The operation performed when the conditions are met and can be regarded as a unit of calculation or transaction processing. After the action is completed, it can be transferred to a new state, it can still be in the original state, or it can be terminated.

4. Substate: The state after the present state transition. When different actions occur and different conditions are generated, a state may transition to a different substate. Once the transition is completed, it becomes the present state.

As shown in **Figure 11**, a task can be represented by a state transition diagram.

SMACH [13–15], which refers to "State Machine," is a powerful and scalable Python-based library for hierarchical state machines. The SMACH library does not depend on ROS and can be used in any Python project. The *executive_smach stack,* however, provides very nice integration with ROS, including smooth *actionlib* integration and a powerful SMACH viewer to visualize and introspect state machines. The SMACH core library is lightweight and mainly provides two interfaces: State and Container.

State: The state represents the state being executed. Each state has some potential outputs. The State class outputs the result by implementing the blocking function *execute()*.

Container: The container is a collection of one or more states that enforces some strategy. The simplest container is a State machine. A SMACH state machine can be viewed as a state flow graph, where each node is an execution state (the robot is performing a certain action), and the edges connecting the nodes represent transitions between states. The State machine itself can also be regarded as a state and has its own output, so they can be combined in layers to complete a complex task.

Figure 12 shows an example state machine [16].

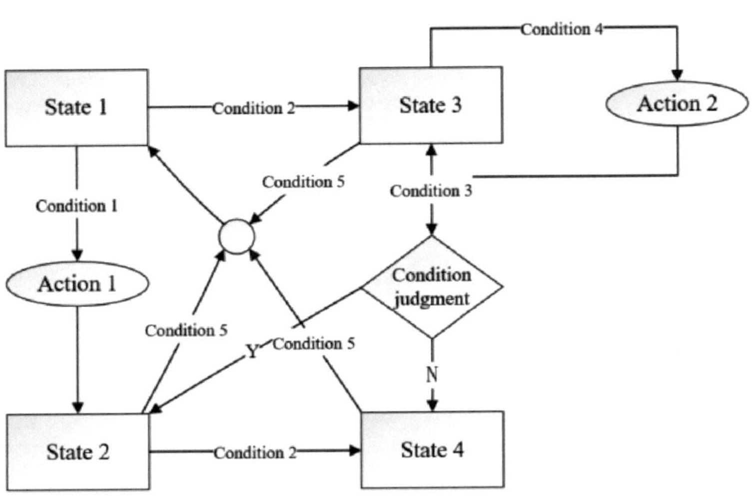

Figure 11.
Finite state machine.

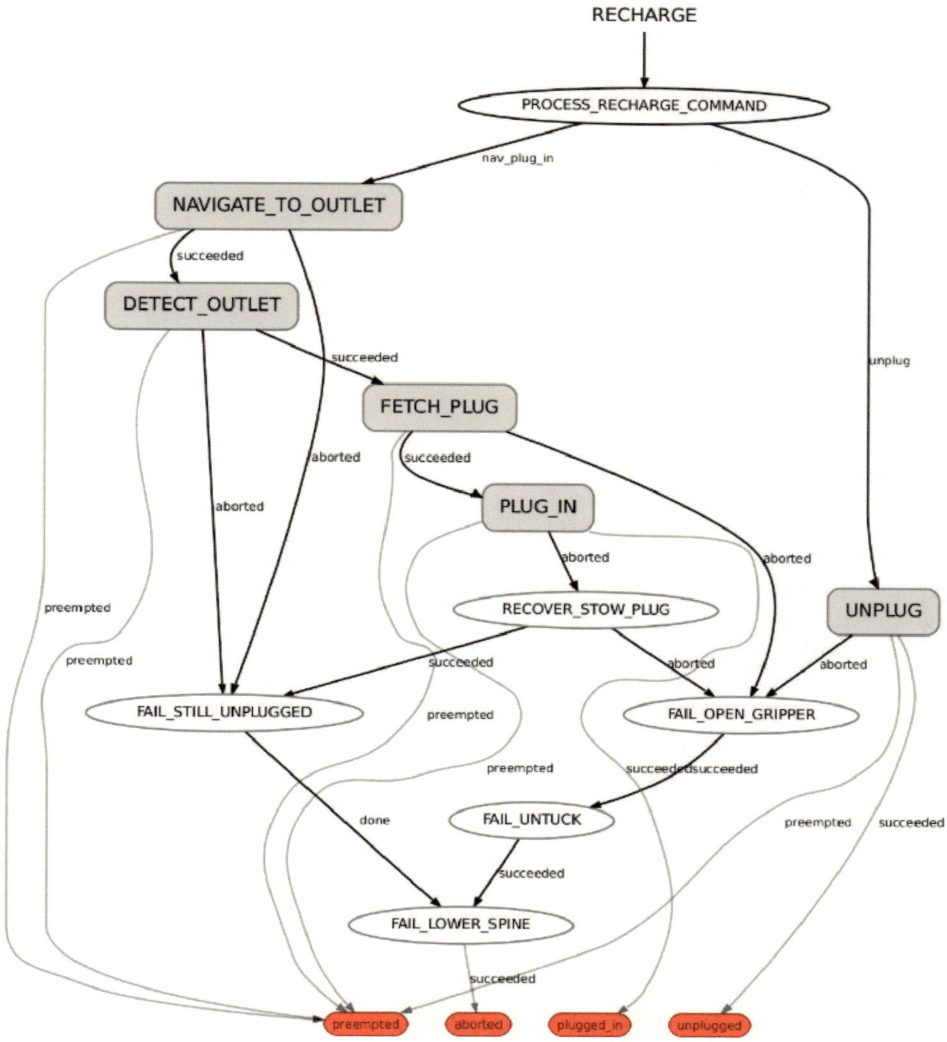

Figure 12.
SMACH state machine example.

SMACH uses action files to define communication protocols between different states. The structure of the action file is simple and clear, as shown in **Figure 13**. Three data definition areas are separated by three underscores. The first area defines the message format of the request, and the middle area defines the returned result (result message format, the bottom area defines the intermediate information feedback) message format. Each area can contain multiple data type, and the system will automatically compile the action file into three message files during the compilation process, so the message format for communication between states is actually the message format provided by ROS.

SMACH provides a general state type to support invocation while providing a special state class *SimpleActionState* as a proxy for *actionlib.* During the construction of the *SimpleActionState* object, the corresponding *actionlib* client is started by default. The user can define a goal in the constructor and create a callback function to process the data returned by the *actionlib* server.

```
# Define the goal
trajectory_msgs/JointTrajectoryPoint target_joint
bool arm
---
# Define the result
int32 succeed
---
# Define a feedback message
int32 percent_complete
```

Figure 13.
File format of action.

3.4.2 Hierarchical concurrent state machine design

For the picking robot to perform a continuous picking task, we can use the state transition process to describe it. The detailed description of the state transition is as follows: first enter the startup state, start the picking robot platform, wait for the initialization of each component, and perform a startup self-test. If any component fails to initialize and is in a fault state, the startup fails, and then the system is placed in the error state, stopping working and the task ends.

If the startup is successful, the platform moves into the state, the mobile platform moves to the first picking point, and data collection state and the tomato scanning state are started at the same time. The data collection state collects tomato information in the current status and uploads it to the database of the server; the tomato scan state checks whether there are tomatoes suitable for picking in the viewing area. If not, restart the platform moving state and move the robot to the next picking point; if so, first analyze all the tomato position information and pass the spatial attitude information of the first target tomato to the kinematics solution state according to the predetermined rules, and then program performs kinematics calculation and motion planning. If the tomato is unreachable, the information of the next target tomato is passed to the kinematics solution state, and so on until the last tomato is reached. If it is determined that the tomato is unreachable, the platform is moved to the next picking point. If it is judged that the tomato is reachable, the calculated right arm motion information is transmitted to the robot motion state, and this state sends the trajectory information of the right arm motion to the lower computer and simultaneously detects the joint motion position information during the execution of the lower computer. After the right arm moves to the target position, it enters the suction state. The system starts the suction device to fix the tomatoes and move them to a suitable position, which is convenient for the left arm to cut hands. The kinematics solution state is started again, and the left arm motion information is solved and transmitted to the robot motion state to control the left arm to move near the target. The start of the visual servoing state is close to the target tomato precisely, and the cutting state is started after the arrival, the pneumatic shear transposition of the left arm is started, and the tomato is cut. After completing a tomato pick, pick the next goal planned. Repeat until the last picking point.

According to the task execution process and state transition process described above, the designed state transition diagram is shown in **Figure 14**.

In SMACH, we use *SimpleActionState* to directly simulate the server side of *actionlib* and define a state machine with 10 states to control the robot to complete a comprehensive picking job.

1. DO_START: start state.
2. DO_MOVEBASE: mobile platform mobile status.
3. DO_TOMATO_SCAN: tomato scanning status.
4. DO_SPATIAL_TEMPORAL: data collection status.
5. DO_KINEMATICS: state of kinematics solution.
6. DO_MOVE_ROBOT: left and right arm movement status.
7. DO_MOVE_ENDEFFECTOR: end effector status.
8. DO_ERROR: fault status.
9. DO_STOP: emergency stop status.
10. DO_VISUALSERVOING: visual servoing status.

3.5 Major software node design

Based on the software framework design and operation requirements of the dual-arm picking robot, based on the functional division of each part, a vision module, an eye-hand coordination module, and a task planning module are mainly designed and installed in the industrial computer. Each module further refines

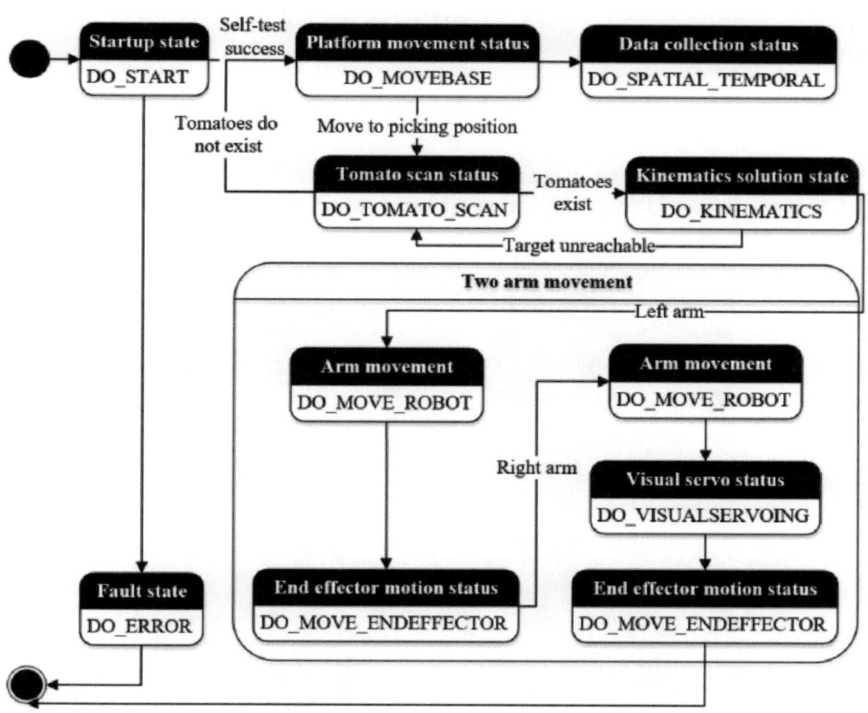

Figure 14.
Continuous picking status flow.

the functions and can be divided into functional nodes. Nodes are the minimum functional modules in ROS, which regard as the ultimate goal of the design.

3.5.1 Vision module node

In the dual-arm acquisition robot eye-hand system, the main component of the "eye" is the camera, including a binocular camera mounted on the robot's head and a monocular camera mounted on the arm. Based on the ROS framework, we designed three nodes to complete the environment perception function: binocular camera image acquisition node (*dual_eye_image_capture*), monocular camera image acquisition node (*single_eye_image_capture*), and image processing node (*image_processing*). The actual recognition effect is shown in **Figure 15**. There are three valid tomatoes in the image. The system recognizes all tomatoes and marks the positions of the tomatoes in the picture that need to be picked first according to the rules.

The binocular camera acquisition node uses the two original images collected by the left and right sensors of the Bumblebee2 camera to finally generate five images: left and right eye corrected color images, left and right eye corrected gray images, and 3D point clouds (**Figure 16**).

The collection process of the binocular camera is shown in **Figure 17**. The camera's original data are read, and the data are packaged into a Bell template image; then, three color information is extracted from the Bell template image and assembled into the original color image. The eye image data are used to obtain corrected left and right eye color images and grayscale images. Next, the left and right eye images are used for stereo matching through the principle of triangulation to generate a 3D point cloud. In the end, all the five images generated were published, and the algorithm used in the image acquisition process was provided by the camera SDK.

Based on the above process, we designed the binocular collection program UML as shown in **Figure 18**.

3.5.1.1 Monocular camera image acquisition node

For monocular vision, we use a Daheng Mercury series industrial camera MER-500-7UC, which uses USB2.0 digital interface and provides free SDK and secondary

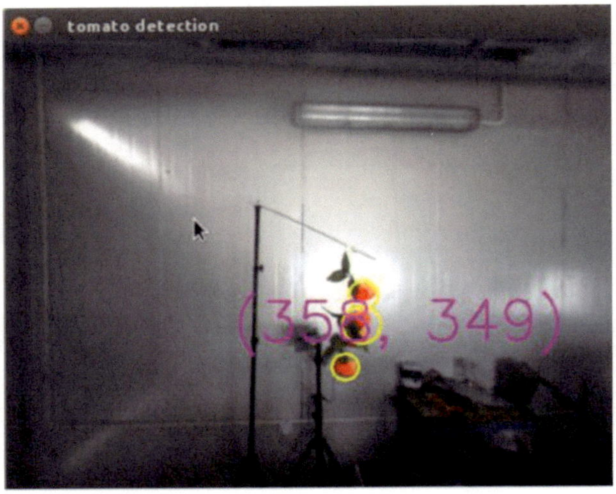

Figure 15.
Tomato identification interface.

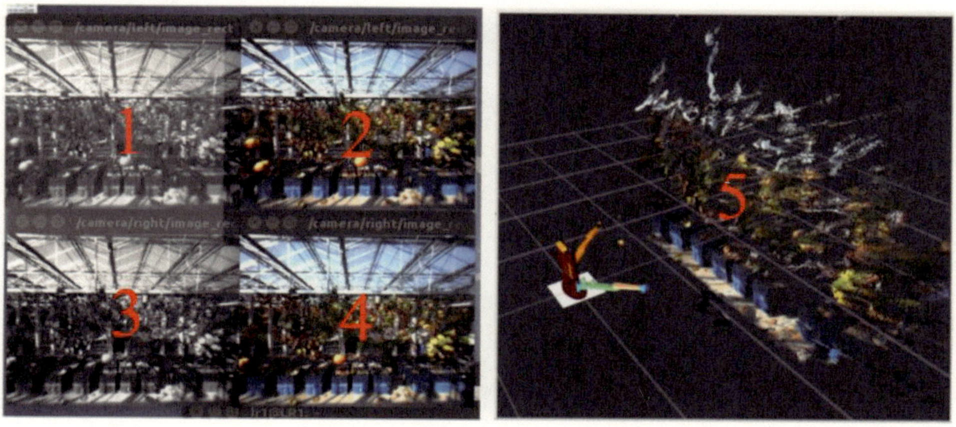

Figure 16.
Five pictures generated by Bumblebee2 camera. 1. Gray image of left eye after correction. 2. Color image of left eye after correction. 3. Gray image of right eye after correction. 4. Color image of right eye after correction. 5. Point cloud image.

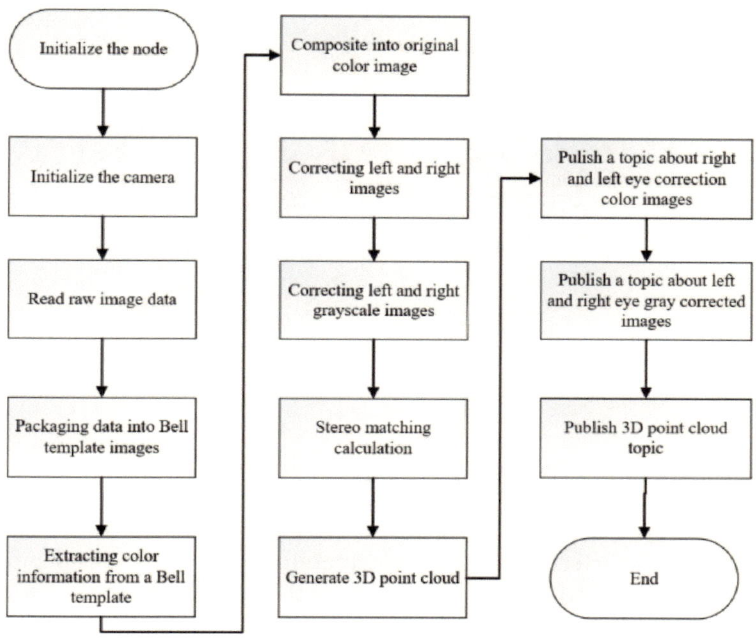

Figure 17.
Binocular image acquisition flow chart.

development example source code under windows platform and Linux platform. We use the *usb_camera* package provided by ROS to collect monocular images, as shown in **Figure 19**.

3.5.1.2 Image processing node

The image processing node receives the collected planar image and point cloud image and provides different image processing function interfaces according to different business requirements. In the current task requirements, image processing nodes are required to complete the accurate two-dimensional recognition

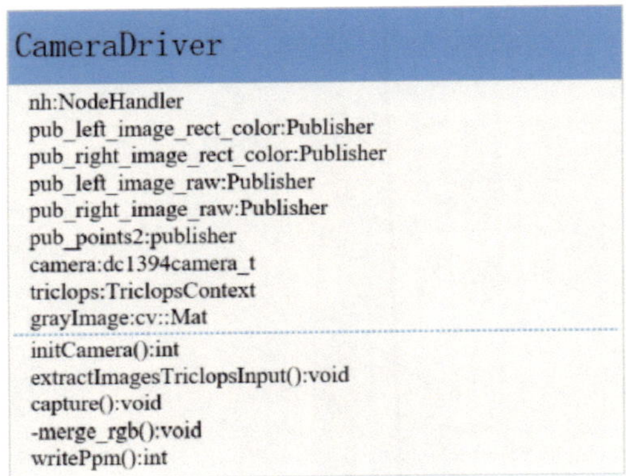

Figure 18.
Binocular acquisition node UML design.

Figure 19.
Image captured by monocular camera.

and accurate three-dimensional positioning of tomatoes. Therefore, the two-dimensional image dataset of the scene and the three-dimensional point cloud data are also required. Next, we introduce us from two directions. Image processing node design: first is the architecture design and functional flow of the image processing node as a functional interface, and the second is the specific implementation of related image processing algorithms.

The entire software system is based on the C/S model architecture, using the *actionlib* function package provided by ROS, with the task planning node as the server, and requesting computing resources from the client of each functional unit. Image processing nodes are no exception. After receiving the image processing instructions and image data, the instructions are parsed to clarify the functional requirements, and then the required image data are extracted, input into the algorithm function for processing, and the results are finally returned to the server.

The specific processing flow is shown in **Figure 20**. After initializing the node and *actionlib* server, start the service, wait for the goal sent by the client, and subscribe to the processing function. After receiving the instruction, analyze the source of the instruction. If the instruction originates from the spatial positioning of tomatoes, the processing steps are: first, use the tomato recognition algorithm based on image feature fusion to identify all tomatoes in the right eye image space of the binocular camera. If there are no tomatoes, return the results; if tomatoes

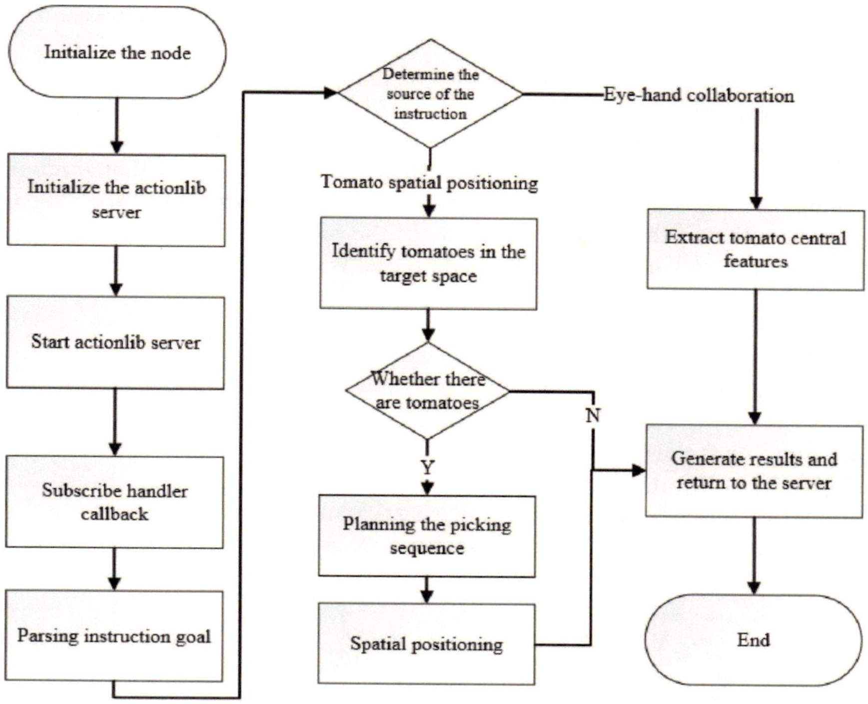

Figure 20.
Image processing node flow chart.

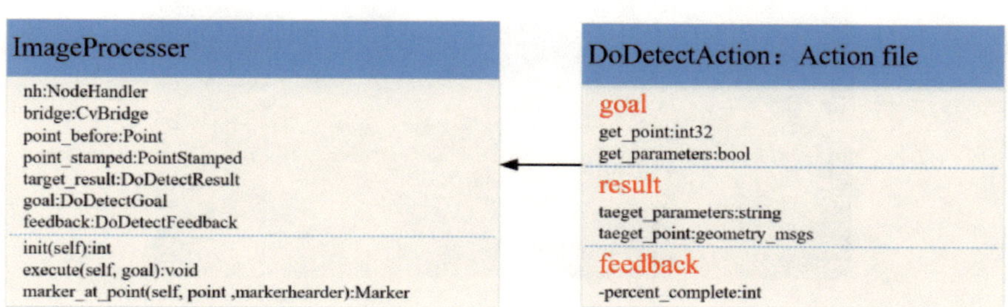

Figure 21.
Image processing node UML design.

are detected, plan the picking order. The rule is from bottom to top, left to right, and calculate the spatial position of the pick point, and finally return the result to the client. If the object recognition result triggers the harvesting task, the image collected by the monocular camera is used to extract the central image feature of the tomato.

According to the above process, the design program UML is shown in **Figure 21**.

3.5.2 Eye-hand coordination module node

3.5.2.1 Eye-hand collaboration process design

Our solution uses an eye-in-hand vision servo solution to achieve eye-hand coordination, as shown in **Figure 22**. The picking robot obtains the image information of the target fruit through a monocular camera installed on the picking hand, extracts the position information of the tomato features in the two-dimensional image, and makes a difference from the expected position information. The difference is used as the input of the visual servo control algorithm and then calculate the control output in real time, that is, the speed vector of the end effector, and then integrate this speed vector with time to calculate the next point that needs to reach the target position. Cycle back and forth to get a trajectory that gradually approaches the target position. The eye-hand correspondence is converted into the amount of motion of the joint, and the end of the robot arm moves accordingly to approach the target. The implementation process is shown in **Figure 23.**

3.5.2.2 ViSP

ViSP [17] is an open source visual servo framework developed and maintained by the Lagadic team of the French National Institute of Information and Automation. It has the characteristics of hardware independence, scalability, and portability. In addition, ViSP also provides a complete library of basic functions, which can be combined with a variety of visual feature libraries; it also provides a simulation environment and interfaces with various hardware. Based on ViSP, we

Figure 22.
Visual servo program of eye-in-hand.

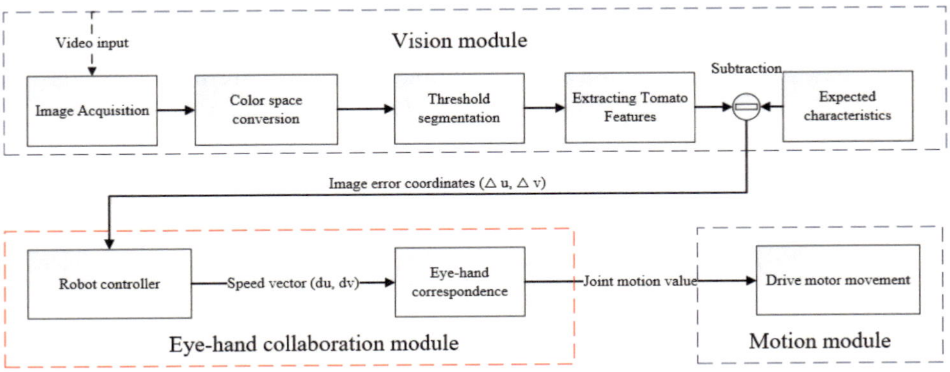

Figure 23.
Eye-hand coordination process.

can complete functions such as visual tracking, fiducial marking, two-dimensional contour tracking, pose estimation, and so on. The goal of ViSP is to provide developers with a tool for rapid development of visual servo functions. The software framework of ViSP is shown in **Figure 24**. The entire framework is divided into three modules: one module provides vision models, vision servo control algorithms, and robot controller interfaces; the second module provides image processing algorithms, tracking algorithms and other machine vision algorithms; and the last module is a visualization module that provides a simulation and visual environment. All these features make ViSP very suitable for use as a core part of our module.

3.5.2.3 Eye-hand collaboration module node design

The complete flowchart of eye-hand coordination is shown in **Figure 25**. After the node is initialized, the system initializes and starts the '*/visual_servo' actionlib* service and subscribes to *execute()* to wait for the client to be awakened. After receiving the service request, start the visual servo loop. In the loop, program request the feature position of the tomato image from the vision module and make a difference from the expected position. If the difference exceeds the threshold Δs (Δs=2mm), the program will obtain the camera parameters, initialize the control model, and call the ViSP library function *vpServo()* to calculate the control output speed vector. Then, program integrates the velocity vector with time (t = 1s), motion module controls robot to move to the output position, and requests the tomato image feature position from the vision module again, then makes a difference with the desired position, and loops back and forth until the target image feature. The difference between the position and the desired image feature position is less than the threshold Δs, the visual servo loop is ended, and our execution result is returned.

Figure 26 shows the design of the eye-hand coordination node class. There are mainly two classes. The *VisualServoCycleNode* class is responsible for the loop and interaction with other modules. The *VisualServoControlNode* module is responsible for controlling the operation of the algorithm.

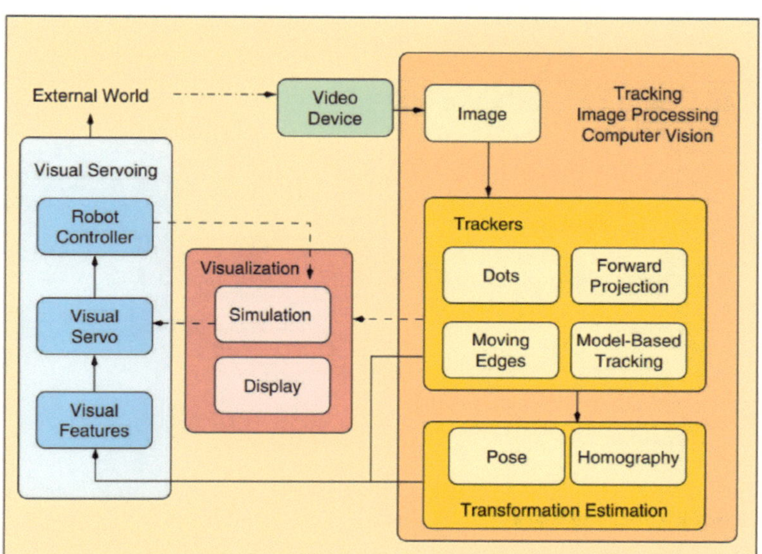

Figure 24.
ViSP software architecture.

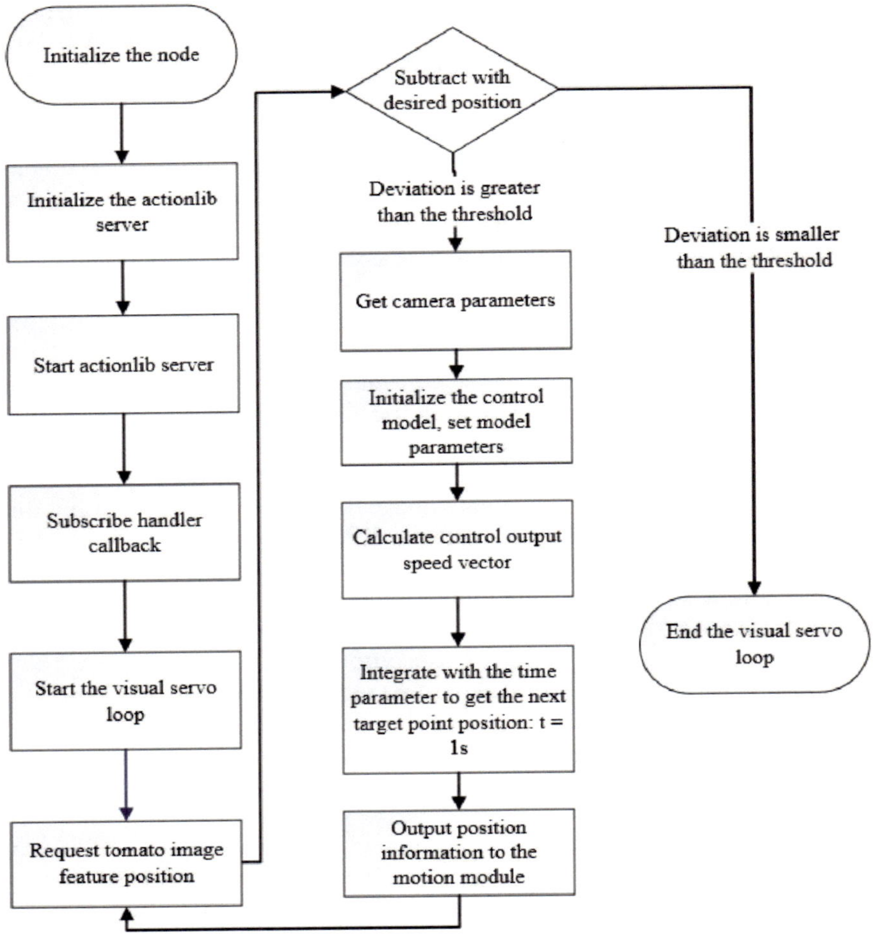

Figure 25.
Eye-hand collaboration node flow chart.

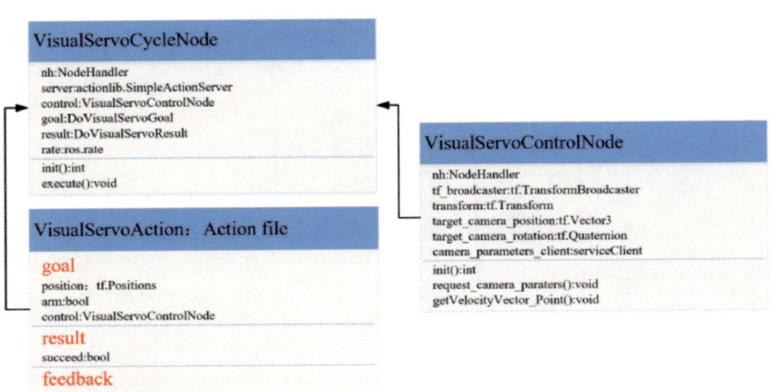

Figure 26.
Eye-hand coordination node class design UML diagram.

3.5.3 Task planning module node

The task planning module mainly completes the design and implementation of a layered concurrent state machine for one pick, as shown in **Figure 27**:

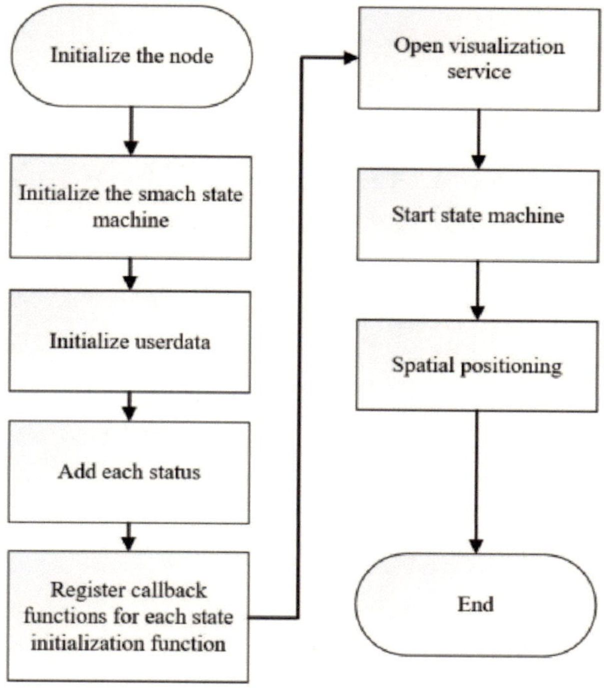

Figure 27.
Task planning node flow chart.

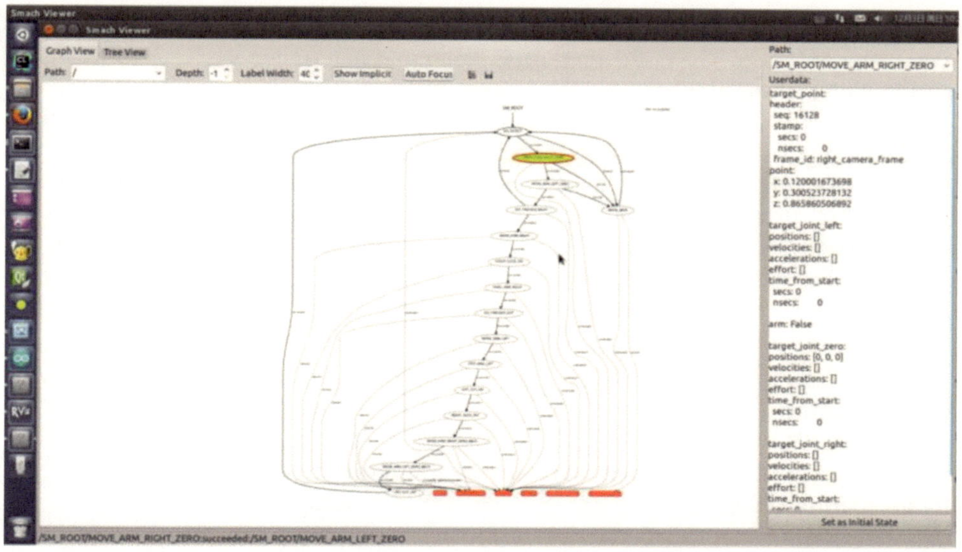

Figure 28.
FSM in SMACH_viewer.

First we initialize the node, state machine, and user intermediate data, and then add the transformation relationship between the states of each state machine according to the state transition of the task design. Use the transition keyword to control the transition from the current state to the secondary state. At the same time, since each state is *SimpleActionState*, each state implements an *actionlib* client by default. You need to add an initialization function and a callback function *callback()* for each

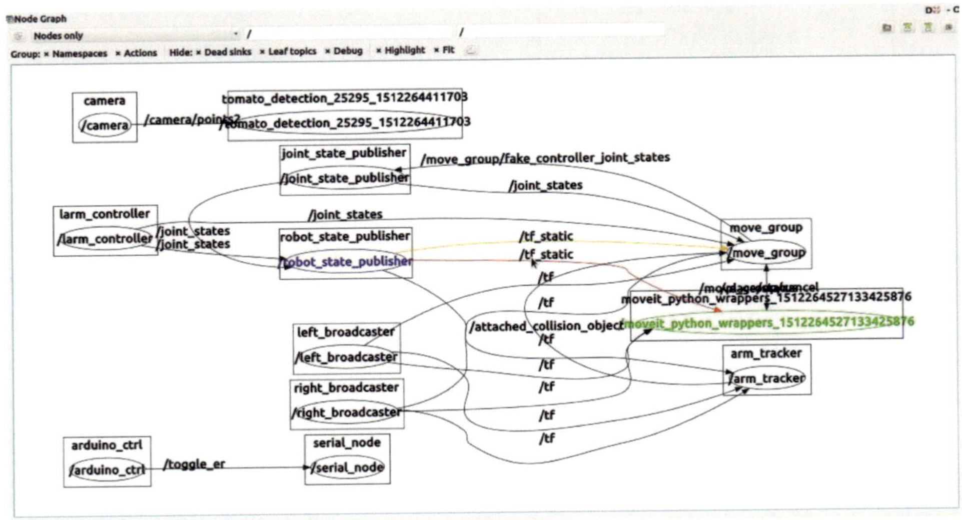

Figure 29.
System function node diagram.

state. Start a state machine visualization service *IntrospectionServe*r in the node, so that we can view the state transition diagram in *SMACH_viewer* and can monitor the state transition in real time. The data details of each state are shown in **Figure 28**.

3.5.4 System node diagram

Figure 29 shows that the running node diagram after all ROS nodes in the system is turned on. The node diagram is generated using the *rqt_graph* command. Each rectangular box represents a topic. The oval box represents a node, and the arrowed lines represent the subscription relationship between each other. Visualization of the node diagram makes the system architecture intuitive.

Since most of the eye-hand coordination and motion control are concurrent, the fluency of multitasks is verified under two plant factories and three greenhouses with different fruit status and illumination variations. The experimental results show that if total number of targets within the visual field is not more than three, the average picking time is less than 35 s.

4. Conclusion

The contribution of this research mainly orients around the software engineering for manipulating the complex robot behavior. Although service robot leverages ROS for rapid development, classical tasks such as eye-hand coordination and continuous operation in an open scenario have not been systematically addressed. In this chapter, we advocate that if the complex robot behavior can be structured, then they can be modeled as Finite State Machines (FSM), and a "Sense Plan Act" (SPA) process can be implemented with a formal software architecture. Meanwhile, we demonstrate that ViSP and SMACH in ROS are beneficial frameworks for developing a dual-arm robot for autonomously harvesting the fruits in plant factory, which embodies the complexity of multi-task planning and scheduling in natural scenes. The experimental results show that the software engineering paradigm effectively improves the system reliability and scalability of the dual-arm harvesting robot.

Acknowledgements

This work was supported by the National Natural Science Foundation of China (No. 51775333) and the Scientific Research Program of Shanghai Science and Technology Commission (No. 18391901000).

Author details

Chengliang Liu, Liang Gong* and Wei Zhang
School of Mechanical Engineering, Shanghai Jiao Tong University, Shanghai, China

*Address all correspondence to: gongliang_mi@sjtu.edu.cn

References

[1] Song J, Zhang T, Xu L, et al. Research actuality and prospect of picking robot for fruits and vegetables. Transactions of the Chinese Society for Agricultural Machinery. 2006;**37**:158-162

[2] Kondo N et al. Fruit harvesting robots in Japan. Advances in Space Research. 1996;**18**:181-184. DOI: 10.1016/0273-1177(95)00806-P

[3] Zhao Y, Wu C, Hu X, et al. Research progress and problems of agricultural robot. Transactions of the Chinese Society of Agricultural Engineering. 2003;**19**:20-24

[4] Tanigaki K et al. Cherry-harvesting robot. Computers and Electronics in Agriculture. 2008;**63**:65-72. DOI: 10.1016/j.compag.2008.01.018

[5] Hemming J et al. A robot for harvesting sweet-pepper in greenhouses. In: Proceedings International Conference of Agricultural Engineering. Zurich; 06-10 July 2014

[6] Taqi F et al. A cherry-tomato harvesting robot. In: 18th International Conference on Advanced Robotics (ICAR); 10-12 July 2017. Hong Kong. New York: IEEE; 2017. pp. 463-468

[7] Nagata M et al. Studies on automatic sorting system for strawberry (part 3) development of sorting system using image processing. Journal of the Japanese Society of Agricultural Machinery. 1997;**59**:43-48

[8] Zhaoxiang L, GANG L. Apple maturity discrimination and positioning system in an apple harvesting robot. New Zealand Journal of Agricultural Research. 2007;**50**:1103-1113. DOI: 10.1080/00288230709510392

[9] Guo F et al. Fruit detachment and classification method for strawberry harvesting robot. International Journal of Advanced Robotic Systems. 2008;**5**(1):41-48. DOI: 10.5772/5662

[10] Wang X. Study on information acquisition and path planning of greenhouse tomato harvesting robot for selective harvesting operations [thesis]. Zhenjiang: Jiangsu University; 2012

[11] Ling X, Zhao Y, Gong L, Liu C, Wang T. Dual-arm cooperation and implementing for robotic harvesting tomato using binocular vision. Robotics and Autonomous Systems. 2019;**114**(4):134-143

[12] Foote T. tf: The transform library. In: IEEE Conference on Technologies for Practical Robot Applications (TePRA); 22-23 April 2013. Woburn. New York: IEEE; 2013. pp. 1-6

[13] Bohren J, Cousins S. The SMACH high-level executive [ROS news]. IEEE Robotics and Automation Magazine. 2010;**17**(4):18-20. DOI: 10.1109/MRA.2010.938836

[14] Mcgann C et al. Model-Based, Hierarchical Control of a Mobile Manipulation Platform. Thessaloniki, Greece: ICAPS Workshop Planning and Plan Execution for Real-World Systems; 2009

[15] Meeussen W et al. Autonomous door opening and plugging in with a personal robot. In: IEEE International Conference on Robotics and Automation; 3-7 May 2010. Anchorage. New York: IEEE; 2010. pp. 729-736

[16] Joseph H. Getting Started with Smach [Internet]. 2018. Available from: https://wiki.ros.org/smach/Tutorials/Getting%20Started [Accessed: 21 March 2020]

[17] Marchand E et al. ViSP for visual servoing: A generic software platform with a wide class of robot control skills. IEEE Robotics and Automation Magazine. 2005;**12**(4):40-52. DOI: 10.1109/MRA.2005.1577023

Chapter 8

Robotization and Welfare Trends in Future

Belma Kencebay

Abstract

There are concerns over the present and possible future impact of new advancements like robots and artificial intelligence on welfare. Experts from different fields including science and business have been concentrating on how new developments may affect the job market, and more broadly how new advancements will influence the society. It would be easy to get support for the use of robots for the tasks which are too difficult or too dangerous for humans. What is the capital owners' focus at that point? What are the economic and social consequences of robotization? In this chapter, literature review including the recent thoughts on how developments in robotics may cause major changes in welfare distribution and revolutionary economic changes is presented.

Keywords: robots, personal economy, robotics, economic effect of robotics, welfare trends, unconditional basic income (UBI) , guaranteed minimum income (GMI), industry 4.0, skill premia, skill-biased technical change (SBTC), polarization

1. Introduction

The new World Robotics report indicates that more than 2.4 million robots are working in manufacturing lines. The robot sales amount is around 16.5 billion USD. As indicated by The International Federation of Robotics (IFR) public statement on Feb 19, 2020 from 2020 to 2022 right around 2 million new units of robots are relied upon to be introduced in industrial facilities around the globe [1].

Robotics technology developed majorly from 1990 to 2000s, especially with an increase in the number of industrial robots in the United States and Western Europe from 1993 to 2007. In the United States, the rise numbered to one new robot per 1000 workers, and in Western Europe to 1.6 new robots per 1000 workers. The automotive industry utilizes the major part of it by 38% of existing robots, then the electronics industry follows it by 15% and the plastics and chemicals industry follows it by 10% and lastly metal products industry by 7%. Acemoglu theoretically found that robots can decrease employment and wages and that their local effects can be evaluated utilizing variety in exposure to robots—defined from industry-level advances in robotics and local industry employment. It is assessed that the relevant field most threatened by robots after 1990 does not exhibit any differential trends before then, and robots' effect is separated from other capital and technologies. One more robot per 1000 workers decreases the employment-to-population ratio by 0.2% and wages by 0.42% [2].

The 47% of laborers in USA will be exposed to risk of losing their employment as indicated by Frey and Osborne [3] study including characterization of

702 occupations dependent on its degree of lack of protection against automation. Other supporting report has been published by the World Bank assesses that 57% of professions in the OECD could be robotized all through the upcoming two decades [4, 5].

Frey and Osborne [3] foresees a shortening of the current trend toward labor market polarization; computerization is mainly related to low skills and low paid professions. Their findings imply that, as technology advances, low experienced labors will redistribute assignments that are not delicate to computerization. Frey and Osborne [3] divide between high, medium, and low risk professions based on the possibility of computerization. According to the study, about 47% of total US employment is in the high risk category [6].

Expanding on the evaluation performed by Frey et al., other investigation assesses the effect of robotization for the OECD including 32 nations that have partaken in the survey of adult skills. It infers that 14% of occupations in OECD would be profoundly controlled by automation (i.e., likelihood of computerization of over 70%). More than 66 million employees from 32 different countries included in Frey's study. According to the study, another 32% of occupations have a danger of somewhere in the range of 50–70% highlighting the chance of huge change in the manner these jobs are fulfilled due to robotization. For example, a critical portion of errands, yet not all, could be robotized, changing the expertise prerequisites for the employments [7–9].

We may see the development of new instruction programs, especially an expansion in computer generated reality gaming-based training and particularly in coding. Be that as it may, this will be tempered by the rise of AI/apply autonomy moving into the information segment, which can possibly prompt the disappearing of expert class, non-administrative desk occupations. Simultaneously, mechanical autonomy will move forcefully into sensor-based, world-route occupations like transportation. We may require a crucial reshaping of our economy and may not give educating/training individuals for occupations that are just not going to be there. And also recent studies found that high levels of anxiety about robotization and automation indicate the broad concerns about their outcomes [7].

It turns out a considerable amount of vulnerability and contention remains. In this chapter, the recent thoughts on how AI and robotics may cause major changes in welfare distribution and revolutionary employment changes will be summarized by a literature review.

2. Literature review

In the study, we will look at two main topics, the first is literature about robotization and its effect on employment and the second is possible future welfare distribution models.

2.1 Robotization and unemployment

In this area, the academics are divided into two poles, one pole assumes that the firm-level adoption of robots causes a decrease in labor shares, rises in added value and productivity, and reduces the share of production workers. Especially in industrial sectors which are specialized in routine activities are more likely to face substantial decreases in the labor share. Some of the studies theorize that new changes in automation, robotics, and artificial intelligence (AI) might pave the way for broad unemployment. All types of occupations from lawyers to truck drivers will be permanently tumbled. Companies will be forced to change or expire. The latest economic signals also reflect this trend; less people are employed, and wages are declining even as productiveness and earnings rise. The list below includes the academicians whose articles associated the robotization and automation to increase in unemployment.

Robotization effects on employment	
Studies	**Dimensions/ideas**
Acemoglu [10]	Skill premia/skill-biased technical change (SBTC)
Acemoglu and Restrepo [2]	SBTC/polarization/ICT
Acemoglu and Restrepo [11]	The displacement effect
Acemoglu and Restrepo [2]	Robotization/unemployment
Acemoglu et al. [12]	Robotization/unemployment
Autor and Dorn [13]	SBTC/polarization/ICT
Autor et al. [14]	SBTC/polarization/ICT
Autor [15]	Polarization
Autor et al. [16]	SBTC/polarization/ICT
Autor et al. [14]	SBTC/polarization/ICT
Berman et al. [17]	SBTC/polarization/ICT
Brynjolfsson and McAfee [18]	Robotization/unemployment/structural changes
Brynjolfsson et al. [10]	The "capital deeping" with robotization
Dao et al. [19]	SBTC/routinized job industry based
Frey and Osborne [3]	Robotization/unemployment
Goos and Manning [20]	SBTC/polarization/ICT
Goos et al. [21]	SBTC/polarization/ICT
Graetz and Michaels [22]	SBTC/polarization/ICT
Gregg and Manning [23]	Wage inequality/SBTC
Huang and Rust [24]	Artificial intelligence (AI)/unemployment
Krusell et al. [25]	SBTC/polarization/ICT
Goos et al. [21]	Polarization
Michaels et al. [26]	SBTC/polarization/ICT
Nedelkoska and Quintini [9]	Robotization/unemployment
Sachs and Kotlikoff [27]; Benzel et al. [28]	SBTC/polarization/ICT
Sachs and Kotlikoff [27]; Benzell et al. [28]	Robotization/unemployment/structural changes
Sachs et al. [9]	Robotization/unemployment
Susskind and Susskind [29]	Robotization/unemployment
Van Reenen [30]	Wage inequality/SBTC/polarization
Wolter et al. [31]	5-step scenario transformation to Industry 4.0

As can be seen above, the dimensions that are studied related with robotization and employment, is listed as skill-biased technical change, polarization, wage inequality, and skill premia. Skill premia can be described as the relative wage of high-skilled workers to low-skilled workers and according to Brynjolfsson and McAfee [18, 32], it has risen over most of the second half of the last century, despite large increases in the supply of high-skilled workers. Acemoglu [10] developed a model to explore how skill premia changes by time and between countries and how it work on this framework to see the effect of foreign commerce on wage inequality. Skill premia is regulated by technology and the supply of skills. Skill-biased technical change (SBTC) can be described as a transfer in the manufacturing technology that prefer competent over incompetent employees by increasing its relative productivity and, therefore, its relative demand. In the literature, SBTC has been studied for several years in that the productivity of more competent workers has risen more faster than that of incompetent workers. This SBTC has been getting quite high recognition due to its relation with increasing wage inequality. Gregg and Manning [23] concluded that when workers are paid according to their productiveness, alteration in productivity reflects revision in wages, and the labor market position of the incompetent workers will continue to fall apart and disappear. Goos et al. [21] studied the job polarization (JP) within-industry and between-industry components empirically and concluded that the employment design in Western Europe has been polarizing with increasing employment proportions for well-paid occupations and supervisors and also flat-salaried service employees and increasing unemployment shares of manufacturing and routine office employees. The JP is a kind of technological change, favoring toward exchanging employees who are doing routine tasks and then the tasks are offshored and both polarization and offshoring create a decline in the demand for average-skilled workers relative to competent workers and incompetent occupations.

Acemoglu [2] contemplated a few sources including a survey by the Ministry of Industry, information provided by French robot suppliers in addition to list of clients, customs data on imports of industrial robots by firm, and the French fiscal files and developed theoretical expectations about adoption of robots for French assembling companies and examine the level ramifications of robot usage. Out of 55,390 firms in their study, 598 have received robots somewhere in the range of 2010 and 2015. Their study indicated that how companies alter their manufacturing system, recruitment, work force portions, and productiveness as they embrace automation technologies that can help us to better understand the wide-ranging effects of automation. Nevertheless, company-level effects do not correspond to the overall impact of automation because firms that adopt such technologies decline the costs and broaden at the expense of competitors. Acemoglu [2] predicts that French manufacturing firms that embrace robots not only decrease their work force portion and percentage of manufacturing laborer and rise their productivity but also expand their operations and employment. Yet, this is more than offset by significant declines in their competitors' employment. Generally speaking, despite the fact that organizations receiving robots extend their work, the market-level ramifications of robot appropriation are not positive. They also declare that robot acceptance commits to the decrease in the production work force portion by decreasing the covariance between firm-level value added and labor share, and this is because adopters are large and enlarge further as they observe sizable relative decreases in their work force portions. The 20% rise in robot adopting is resulted with a 3.2% increase in industry unemployment [12].

Acemoglu and Restrepo started a framework published in 2019, according to the heart of their framework is the fundamental thought that computerization and along these lines AI and mechanical technology supplant laborers in undertakings that they recently performed, and by means of this channel, make an incredible "The displacement effect." As opposed to assumptions in a lot of macroeconomics and financial matters, which keep up that profitability upgrading innovations consistently increment in general work request, the relocation impact can diminish the interest for work, wages, and business. In addition, the "displacement effect" infers that increments in yield per specialist emerging from robotization will not bring about a relative development of the interest for work. The removal effect causes a decoupling of wages and yield per laborer and a decrease in the portion of work in national salary.

Acemoglu et al. [12] at that point feature a few countervailing powers that push against the uprooting impact and may infer that computerization, AI, and apply autonomy could build labor demand. First, the replacement of modest machines for human work makes an "efficiency impact": as the expense of delivering mechanized assignments decays, the economy will grow and expand the interest for work in nonautomated errands. The efficiency effect could show itself as an expansion in the interest for work in similar areas experiencing robotization or as an expansion in the interest for work in nonautomating divisions.

Second, according to Acemoglu et al. [12], "capital aggregation" activated by expanded mechanization (which raises the interest for capital) will likewise raise the interest for work. Third, mechanization does not simply work at the broad edge—supplanting undertakings recently performed by work—yet at the escalated edge also, expanding the efficiency of machines in assignments that were recently robotized. This marvel, which we allude to as "deepening of automation," makes an efficiency effect however no uprooting, and along these lines builds work request. Despite the fact that these countervailing impacts are significant, they are for the most part deficient to incite a "balanced growth path," implying that regardless of whether these effects were incredible, continuous computerization would at present lessen the offer of work in national income (and potentially employment). Acemoglu et al. [12] contend that there is an all the more remarkable countervailing power that expands the interest for work just as the portion of work in national income: the formation of new undertakings, capacities, and exercises in which work has a near preferred position relative to machines. The production of new errands creates a restoration effect straightforwardly counterbalancing the displacement effect [12].

Sachs et al. [33] expect in their study that robots are not to help people in the study, however to supplant them totally. They concluded that the presentation of robots will help profitability in the short term, yet decline wages and utilization over the long haul. Sachs and Kotlikoff [27], expecting that "brilliant machines" supplant youthful and untalented yet favor old and talented work, locate that lone a generational arrangement can make the presentation of robots a gainful situation for the two ages. So also, Sachs et al. [33] contend for government redistribution in this situation to counter the "immiserization" of people in the future. Autor [34] reacts to these alerts by expressing that in these models "the key danger isn't innovation essentially yet misgovernance": it is not an issue of shortage of employments, but instead a distributional issue that robots undoubtedly make human work pointless. He contends that a fitting capital duty can assist with gaining innovative ground a government assistance improving procedure for all gatherings of laborers [27, 28, 34].

Robotization and advanced innovations all the more for the most part will empower little players, including people and little organizations, to attempt venture work that is currently before were done inside greater firms. The development of little and extremely enormous organizations could make a barbell-shaped economy, in which middle-sized organizations could miss out. It is not yet clear whether computerization could elevate rivalry, empowering firms to enter new zones outside their past center organizations and making a developing separation between mechanical pioneers and loafers in each area [35].

The skill premia (the general pay of high-talented specialists to low-gifted laborers) rise up the majority of the second 50% of the only remaining century, regardless of huge increments in the gracefully of high-gifted laborers. The end was that there more likely than not been something like an SBTC which expanded the interest for high-talented laborers considerably more. Berman et al. [17] were among the frontier to examine the wellsprings of the consistently expanding skill premia. In a comparative vein, Krusell et al. [25] modeled an economy with an interdependent between a sort of capital and high-talented laborers. The kind of capital they have at the top of the priority list is Information and Communication Technology (ICT) capital. Krusell et al. [25] report that the cost of ICT capital has been falling for quite a long while. In this way, such a capital-ability interdependency, a fall in the cost of ICT capital will prompt an expanded reception in firms and consequently to an expanded interest for high-talented specialists to work these machines. Michaels et al. [26] affirm these discoveries with more up to date information: Sectors with higher development in ICT likewise had higher increments in the interest for high-gifted specialists and diminishes in the interest for center talented laborers. Spitz-Oener [36] secures that position necessities have been expanding in a similar time, for example, the extent of complex undertakings has been expanding. These adjustments in the assignment structure have additionally raised the interest for aptitudes in the labor market [25].

Dao et al. [19] discover that industrial sectors which had some expertise in routine exercises would in general experience bigger abatements in the labor share. Graetz and Michaels [22] use open data on robot use to measure the effects on labor profitability development, absolute factor productivity development, unit costs, and employment. Their disclosures show that robots increment labor efficiency development and all out profitability development however will in general reduce output price. While there is by all accounts no impact of robots use and complete business, they locate a negative effect of robots on the work portion of low-gifted specialists [22, 37].

In another investigation, Acemoglu and Restrepo [11] center around US nearby work markets. They join information from EU KLEMS and robot use to follow the impacts of expanded presentation to robots on neighborhood work markets from 1970 to 2007. As Graetz and Michaels [22], they find that the appropriation of robots prompts enormous and strong decreases in work and wages [38].

The movement of AI task substitution from lower to higher intellect (mechanical, scientific, instinctive, and sympathetic) brings about unsurprising movements after some time. As indicated by this view, scientific abilities will turn out to be less significant, as AI assumes control over progressively expository errands, giving the "milder" natural and sympathetic aptitudes considerably more significance for administration representatives. In the end, AI will be equipped for performing even the instinctive and compassionate undertakings, which empowers imaginative methods of human-machine incorporation for offering support yet in addition brings about a key risk for human employment [24, 39].

Other than work replacement, the innovative change is relied upon to influence the structure of employment much more than the degree of employment, something that would make a more polarized labor market between profoundly qualified and low-gifted occupations. An outcome, there would be progressively huge pay imbalances between the two posts.

A primary challenge for the eventual fate of work is related to adapting to rising disparity, as innovative change will generate the victors and failures and an expansion of the working poor [5, 21]. The some of studies concludes that innovation might be the single biggest supporter of the expansion in disparity of salary. This emerges on the grounds that organizations embrace advances at an alternate pace and have varying degrees of accomplishment with their AI and automated changes. Simultaneously, the robotization of laborers' exercises for capital drives down the work portion of pay [7, 35].

Until then some studies from pessimistic view are mentioned, but there are some other studies believe that robotization and employment can be complimentary. The optimistic studies are listed below:

Robotization effects on employment	
Studies	**Dimensions/ideas**
Autor [34]	The automation and labor are highly complementary
Arntz et al. [40]	Automation is hard to adapt/massive diversity of tasks/adaptability of jobs in the digital transformation
Autor and Salomons [41]	Own-industry, between-industry, cross-country, and final demand effects
Bresnahan et al. [42]	Skilled labor is complementary with a cluster of three distinct changes at the firm level: information technology, new work organization, and new products and services.
Doms et al. [43]	Skilled labor and ICT are complementary
Kurzweil [44]	Improvements in technology will yield greater opportunities
Mokyr et al. [45]	Innovations will result in unimaginable new occupations
Negroponte [46]	Borderless new era
Song [47]	Technology, will rid markets of inefficiency and propel humanity toward its fullest potential
Wolter et al. [31]	The Economy 4.0 scenario will create 1.5 million new jobs which will not exist
Zeira [48]	Increases capital requirements/highly productive countries will employ robotization

For example, Autor [34] identify in his study that the automation has not clear out a most of jobs instead, automation will substitute for labor—as it is typically intended to do. But, automation will also complement labor, increases output that create higher demand for labor, and responds to changes in labor supply. He believes that some studies lean to overestimate the context of robot backup for labor and forget the robust complementarities between robotization and labor that rise productivity, increase earnings, and boost demand for labor. Autor [34] also argue in his study that the polarization is doubtful to go very far into the anticipated future. Arntz et al. [40] demonstrates that these pessimistic scenarios are exaggerating the portion of automatable jobs by ignoring the massive diversity of tasks within jobs as well as the versatility of jobs in the digital transformation. In order to support their proposal, they used detailed task data and declared that, when taking into accounting the spectrum of tasks within jobs, the robotization risk of US jobs decrease from 38 to 9%. Earlier studies have produced by Doms et al. concluded that based on the skill-biased nature of ICT that indicate skilled labor and ICT are complementary [43].

Autor and Salomons [41] has also concluded that systematic view of four different channels of how robotization may affect the employee market; own-industry effects, between-industry, cross-country effects, and final demand effects. They stated that total factor productivity has negative direct effects on employment but positive indirect effects. In summary, the positive effects dominate and the long term outcome of robotization on employment is positive. They studied 24 OECD economies and stated that while displacing employment in the industries where it originates, automation generates indirect employment growth in customer industries and rise in aggregate demand, finally bringing net employment growth [41].

Another optimistic study is made by Bresnahan et al. [42]; their study concluded that the competent labor is complementary with a group of three separate changes at the company level: information technology, new work organization, and new products and services.

In labor economics field, replacement of human work by AI and robots is fervently talked about. In any case, as indicated by Autor , robotization and mechanical advancement has not prompted the oldness of human work. Indeed, computerization and labor are exceptionally correlative and are partial to representatives that are versatile, ingenious, and arrangements situated [34].

Taking into account the former practices learned since the Industrial Revolution, Mokyr et al. argue that PCs and robots will make anew things and organizations and that these thing progressions will achieve impossible new occupations [45, 49].

3. Potential economic results of robotization and offered welfare models

The financial change following quick robotization can be more difficult than the procedure we have illustrated for various reasons. Generally clear, robotization changes the idea of existing jobs, and the reallocation of laborers from existing employments and undertakings to new ones is a complex and frequently slow procedure. It requires some investment for laborers to find new openings and errands in which they can be profitable, and periods during which laborers are laid off from their current occupations can make a discouraged local or national labor market, further expanding the expenses of modification.

These impacts are noticeable in ongoing examinations that have concentrated on the change of nearby US work markets to negative interest stuns, for example, Autor [15], who study the moderate and exceptionally inadequate change of local labor markets in light of the flood in Chinese exports, Mian and Sufi [50], who explore the ramifications of the breakdown in housing costs on consumption and local employment. Acemoglu and Restrepo [11], who discover work furthermore, wage decreases in territories generally presented to one specific sort of computerization, the presentation of modern robots in assembling. The chronicled record additionally underscores the agonizing idea of the alteration.

The fast presentation of new advancements during the British Industrial Revolution eventually prompted rising work request and wages; however this was simply after an extended time of stale wages, growing destitution, and furthermore, unforgiving day to day environments. During an 80-year time frame reaching out from the start of the Industrial Revolution to the center of the nineteenth century, compensation deteriorated and the work share fell, even as innovative advances and efficiency development were progressing in the British economy, a wonder which Allen [51] names the "Engel's delay." There should in this manner be no assumption that the acclimation to the changed work advertise realized by fast computerization will be a consistent, costless, also, quick procedure. It is maybe telling that wages began developing in the nineteenth-century British economy simply after mass tutoring and different interests in human capital extended the aptitudes of the workforce.

So also, the acclimation to the huge flexibly of work liberated from agribusiness in mid twentieth-century. America may have been enormously supported by the "secondary school development," which expanded the human capital of the new age of American laborers [52]. The powers at work here are probably going to be progressively broad than these models. New errands will in general require new abilities. Be that as it may, to the degree that the workforce does not have those abilities, the change procedure will be hampered. Considerably more forebodingly, if the instructive framework is not doing giving those abilities (and in the event that we are not in any case mindful of the kinds of new aptitudes that will be required to empower interests in them), the change will be significantly hindered [12].

As for example, demographer Joel Kotkin [53] causes to notice the advancement toward neo-feudalism in California. The Golden State, with a populace of 38 million occupants, has regularly been viewed as a model of a "mesocratic culture" in the US with an extremely enormous white collar class. As indicated by late turns of events, four separated classes have been developed: (a) the theocracy of the super-rich, particularly in fund and IT; the scholarly world class, like academics, (b) media experts or public decision-makers; (c) the white collar class of experts and little owners; and (d) the serfs, spoke to by the "working needy individuals" and those subject to endowments and government help.

As indicated by the discoveries of Kotkin, the once adequate white collar class has been damaged and California has gone into a neo-primitive period, while the oligarchs and the scholarly elites has increased more force and the servants has duplicated all over the place.

In future where respectable and generously compensated occupations for everybody are vulnerable to be rebuilt with the increase of robotization, in such a situation, the best option for the government assistance states to counterbalance laborers' vulnerability is to invigorate "minimum income guarantees" so as to compose social citizenship compelling. This is by all accounts a conceivable strategy in develop European government assistance declare that as they have just encountered a panoply of "safety net" arrangements of social security for the workless [54, 55].

In periods of economic shrink, around the 1930s, and between 1970s and early 1980s, Basic Income discussions are higher than in periods of economic growth (for example, growth of capitalism in the 1950 and 1960s) [56].

What is a UBI? The discussion on a UBI is frequently riotous and without exact definitional forms. In numerous occurrences, a UBI is compared with ensured least pay plans, while others characterize a widespread program as one that does not build up any qualification models other than age. We propose a meaning of UBI dependent on three center structure decisions—that it is paid to all, unequivocally, and in cash.

Giving salary backing to jobless residents has been a long-standing duty in government assistance majority rule governments since the occasions were projects to battle the old social hazard, for example, joblessness were executed. The conventional presumption of such approaches was that, after a period had slipped by in looking through a new position, the laborer may turn into an active employee again. This procedure of work modification shows up not to act naturally apparent any longer. As a result of occupation replacement actuated by robotization, conversation about giving cash backing to "replaced" laborers is picking up force in the public eye. However, the thought is to increase transversal social accord.

Scholar holding differing ideological thinking caution about the conceivable broken impacts that giving cash without working could create. After all, entrance to work as the methods for social incorporation is the choice favored by a dominant part of the individuals in (post) modern social orders. Research and studies exhibit that giving money installments to the least fortunate improves their lives and does not increment inefficient spending or apathy [57, 58].

An another aspect of robotization effect on economy and social orders as well. The robotization effect on "personal economy" is examined by Kencebay [4]. It is stated that Anxiety has higher and quite material effect on Personal Economy. The study anticipated that Personal Economy will have stronger effect on "Intention to

Accept for Robots" [4]. The high unemployment rates and declining wages may rise the anxiety, due to the anxiety's material effect on personal economy parallel to that it may effect the level of acceptance of robots. The acceptance of robot for capital owners and work force can be very different due to the separate and different incentives, the potential controversy is clear but has not answered or studied yet.

Wispelaere et al. concluded that three crucial barriers that may block a basic income concept from providing the desired worldwide scope, the first barrier is obtaining a population-wide eligible applicant ensuring adaptation; the second barrier is instituting robust method of payment that arrive all aimed recipients; and the third barriers is creating an adequate control system in a policy framework [59].

A UBI is frequently connected with being a distinct advantage in power redistribution [60]. Such calls may reverberate and intensify previous impression of injustice and imbalances that are crawling into the establishments of implicit understandings [61, 62].

With regards to UBI, the models that are all inclusive and adequate are probably not going to be reasonable, and models that are moderate are not general. The International Labor Organization gauges the worldwide normal expense for UBI, as an extent of total national output, would be 32.7%. Current worldwide normal government use is 33.5% of GDP [57].

The distribution of revenues may not really be inspired by productivity improvements, however, by social and natural standards. From a social point of view, the thought is to reinforce the social texture by inciting a feeling of regular reason, which continues from characteristic assets being a type of investor profit for residents. Current models in Alaska and the Eastern Band of the Cherokee Nation typify this methodology [63]. The Alaska Permanent Fund is intended to distribute oil credit to occupants, while the Eastern Band of the Cherokee Nation is identified with gambling club benefits: since 1997, the innate government has given a segment of its profits to 16,000 grown-up ancestral individuals. The normal yearly profit is around US$4000 per capita, which is dependent upon government tax assessment and split into two installments for every year.

When it comes to the welfare models and unemployment solution, UBI or GMI have been studied besides other than no nation as of now has such a plan set up, and just two have done so incidentally (Mongolia and the Islamic Republic of Iran). Those encounters offer some accommodating bits of knowledge into center inquiries, for example, financing and inflation, while test cases programs are producing data on at least one characterizing highlights of a UBI. However, framework wide issues are to a great extent left unanswered, for example, the relationship to the lowest pay permitted by law, severance pay, or benefits [61]. Absolutely, for governors a adopt of robotization could come with raise aptitudes and advance employment creation. By reconsidering salary backing and social security nets, new government assistance plans are additionally to be modified [8, 64–66].

The concept of UBI has been studied by country based; ranging from the US [67], to Australia [68], to New Zealand [69], to Sweden [70], to the UK [71], to Belgium and the Netherlands, to Finland [72], and to the rest of Europe [73]. All studies' main conclusion is that UBI may be applicable in these areas of authority, but that operation demands both strong and sustainable capital and political discipline. In the following part, we will look at the basic income trails in history [74].

4. Basic income in history

In 1967, the territory of Alaska encountered an unexpected benefit of oil riches when North America's biggest oil hold was found on state-claimed land. The rent sold for a faltering US$900 million. Driven by Governor Jay Hammond, a 1976 change to the legislation enforced the state to store in any event 25% of every year's normal asset incomes in an Alaska Permanent Fund. The reserve incomes are put resources into an investment account, with part of the premium paid every year to occupants as an asset profit. Since 1982, each inhabitant has been qualified for a yearly UCT from the state. People must apply every year, meet the residency standards (be available in Alaska for the former year, with the expectation to remain uncertainly), and have no ongoing genuine criminal conviction. In excess of 90% of the populace regularly gets the profit. Not at all like a genuine UBI, the sum is neither stable nor adequate to address essential issues. The all out profit dissemination has truly added up to half of the reserve's yearly premium (found the middle value of over the past five-year time frame); however, this was as of late decreased to distribute a greater amount of the store's income to the state's enormous shortfall.

The Alaska Permanent Fund profit has pulled in huge encouragement across segment, financial, and political partitions. The program has brought neediness and imbalance levels down to among the country's most minimal and invigorated the economy, creating more than 7000 occupations and US$1.1 billion personal revenue, without inflation or decreasing work. Indeed, ongoing proof shows part-time work has expanded by 17% [75]. Meantime the profit may have not influenced fertility ratios, it has gone about as a gentle magnet for at any rate 12,000 transients.

The Iranian UBI plot was conceived out of more extensive change bundles. In 2008, the legislature reported a lot of clearing changes in energy and food aids. Nonetheless, opposite general belief to imply testing was developing. Subsequently, the focusing on plan was deserted, supplanted rather with uniform widespread money moves with the rich being debilitated from taking an interest. In January 2010, Parliament affirmed the appropriation change bundle and arrangements were propelled to broadcast the changes, set up the bank framework, guarantee general record get to, and diminish swelling in front of the cost expands [76]. On December 19, 2010, the universal cash move was saved in the financial balances of family heads. Simultaneously, local energy and agricultural costs rose by up to multiple times. From the outset, just 80% of families enlisted for the plan, an offer that immediately rose to 96%.

The amount of monthly cash transfer was set at Rls 455,000 (US$40–US$45) per individual—29% of median per capita income, and about multiple times the sum visualized in the endowment change law. The program assimilates about 3% of GDP. The blend of endowment changes and endorses activated swelling rates that disintegrated the exchanges' buying power by 66% of their unique incentive by 2018. All things being equal, the program achieved promising outcomes. For instance, observational examinations discover by and large no negative work gracefully impact (on hours worked and investment); yet the young worked somewhat less (for the most part since they were joining up with advanced education), while administration laborers worked increasingly [76].

At the automatic level, the decision would rely upon the announcement of the issue that UBI is planned to address—for instance, regardless of whether innovative advances will in the long run bring about gigantic net occupation misfortunes—and how well frameworks or a specific program are neutralizing that objective in a given setting [77]. Evaluating the propriety and practicality of UBI requires understanding and working; however, the exchange offs that any program or set of projects face as far as inclusion, progressivity, sufficiency, motivators, costs, financing choices, political economy, and conveyance. None of these parameters have a simple and prearranged result [78, 79]. A UBI is an apparently straightforward thought that includes complex decisions [61].

5. Conclusions

SEO specialist, social media account representative, Uber driver, Airbnb host, influencer, app developer, and drone operator; these jobs were not exist 10 years ago. Can we image today's 5-year-olds do when they are 25? What kind of jobs will disappear, will robotization create new jobs that compensate the job loss? How employment will be affected by robotization? Which welfare model in this new era would be appropriate? Will UBI or GMI be paid? Will UBI or GMI solve the problem of potential unemployment? Have governments' convenient economic structure to pay these fees for long term? What is the global reality and testimony around UBI?

While computerization has verifiably been connected to routine errands including predetermined, regulated actions, and big data method are currently quickly entering regions that can supplant work in a wide assortment of non-routine subjective undertakings. What's more, exceptional robots increase propelled sense and finesse, permitting them to play out a more extensive manual errand region. This is probably going to change the idea of work among industries and professions.

The mankind has been fearing its economic redundancy for long time. Robots and automation are instilling that same fear in leading academics and concerned politicians alike. Some of the official reports or bestseller books may be predicting a dark future. Machines are anticipated to be increasingly taking over tasks currently executed by humans. An permanent job crisis will be inevitable for virtually everyone except the best brains. That's what was announced... in the early 60s. Pessimists believe that artificial intelligence and robotics will eventually turn most of the working population idle. Computers can make medical diagnoses, deliver legal documents, translate texts, teach students, control planes, presenting care provision, the transport and distribution sector, freight handling, and what is left of industrial production with a new wave of automation. The catastrophe is very close. New welfare systems have to be planned or new political views even capitalism may be fall. Optimists believe that robotization may change the welfare system and job market but pace of robotization and new job opportunities may not create dark scenarios.

Supplanting innovations supplant laborers and may cause joblessness. As opposed to the standard system dependent on empowering advances, supplanting innovations can lessen wages. It may cause polarization of employment. Truth be told, employment development can regularly be in zones where wage development is restricted. There is discussion between who feel that we have come up short on thoughts and who imagine that we are at the edge of making endless thriving with PCs creating PCs delivering robots delivering computers. Those who stress over the troublesome impacts of AI and robots, we despite need a lot of proof, yet a portion of the examples we have seen demonstrate there will be a ton of disturbance. As optimistic scenario's main assumption that as new machines supplant work in certain undertakings may happen, and new assignments in which work has comparative improvement will be made but it seems that optimistic and pessimist scenarios have one common thing; routine tasks conducted by low-skill employees probably face with job extinction. Education system and job structure have to be changed in order to keep pace with high speed technological changes. People's skill needs would be changed, the employees who are open to learn and change would be the survival of this race. Governments has a huge responsibility to maintain social equity without disturb the economic balances.

The investigation of the referenced above scientific works gives the motivation to accept that the likely ramifications for business and the social circle related with maturing of a few and the presence of new callings are still hard to be predictable. From one viewpoint, the fast and difficulty procedure of robotization shows that neither relevant education nor creating of new abilities in future will save from unemployment, the material risk, and threat in the labor market. On the other hand, the mechanisms of employment transformation has to be considered, the appearance of jobs in new industries, the probable introduction of technological, financial, legal, and other restrictions on automation and robotization of jobs.

Author details

Belma Kencebay
Near East University, North Cyprus

*Address all correspondence to: belma_bozdemir@hotmail.com

References

[1] Hagele M. Robots conquer the world [turning point]. IEEE Robotics and Automation Magazine. 2016;**23**(1):120-128. DOI: 10.1109/MRA.2015.2512741

[2] Acemoglu D, Restrepo P. Robots and jobs: Evidence from US labor markets. Journal of Political Economy. 2020;**128**(6):2188-2244. DOI: 10.1086/705716

[3] Frey CB, Osborne MA. The Future of Employment: How Susceptible Are Jobs to Computerization? Working Paper. Oxford Martin School; 2013

[4] Kencebay B. User acceptance of driverless vehicles and robots with aspect of personal economy. Journal of Transnational Management. 2019;**24**(4):283-304. DOI: 10.1080/15475778.2019.1664234

[5] Arntz M, Gregory T, Zierahn U. The Risk of Automation for Jobs in OECD Countries: A Comparative Analysis, OECD Social, Employment and Migration Working Papers, No. 189. Paris: OECD Publishing. DOI: 10.1787/1815199X

[6] Frey CB, Osborne MA. The future of employment: How susceptible are jobs to computerisation? Technological Forecasting and Social Change. 2017;**114**:254-280. DOI: 10.1016/j.techfore.2016.08.019

[7] Acemoglu D, Restrepo P. Artificial Intelligence, Automation and Work. Cambridge, MA: National Bureau of Economic Research; 2018. pp. 9-10. DOI: 10.3386/w24196

[8] Moreno L. Robotization and Welfare Scenarios. Instituto de Políticas y Bienes Públicos (IPP) CSIC, Working Paper. 2019-01; 2019

[9] Nedelkoska L, Quintini G. Automation, skills use and training, OECD Social, Employment and Migration Working Papers, No. 202, OECD Publishing; 2018

[10] Brynjolfsson E, McAfee A, Spence M. New world order: Labor, capital, and ideas in the power law economy. Foreign Affairs. 2014;**93**(4):44-53. Retrieved June 17, 2020. Available from: www.jstor.org/stable/24483556

[11] Acemoglu D, Restrepo P. Robots and Jobs: Evidence from US Labor Markets. MIT Department of Economics Working Paper No. 17-04. 2017. Available from SSRN: https://ssrn.com/abstract=2940245. DOI: 10.2139/ssrn.2940245

[12] Acemoglu D, Lelarge C, Restrepo P. Competing with Robots: Firm-Level Evidence from France2020. DOI: 10.3386/w26738

[13] Autor D, Dorn D. This job is "Getting Old": Measuring changes in job opportunities using occupational age structure. The American Economic Review. 2009;**99**(2):45-51. DOI: 10.1257/aer.99.2.45

[14] Autor D, Katz L, Krueger AB. Computing inequality: have computers changed the labor market? Quarterly Journal of Economics. 1998;**113**(4):1169-1213

[15] Autor D. The Polarization of Job Opportunities in the U.S. Labor Market: Implications for Employment and Earnings. Center for American Progress and The Brookings Institution: The Hamilton Project. 2010. Available from: http://www.brookings.edu/research/papers/2010/04/jobs-autor

[16] Autor DH, Katz LF, Kearney MS. The polarization of the US labor market. The American Economic Review. 2006;**96**(2):189-194